Jet Sense: The Philosophy and the Art of Jet Transport Design

Jet Sense: The Philosophy and the Art of Jet Transport Design

ZARIR D. PASTAKIA

SAE INTERNATIONAL®

Warrendale, Pennsylvania, USA

400 Commonwealth Drive
Warrendale, PA 15096-0001 USA
E-mail: CustomerService@sae.org
Phone: 877-606-7323 (inside USA and Canada)
724-776-4970 (outside USA)
Fax: 724-776-0790

Library of Congress Catalog Number 2023952263
http://dx.doi.org/10.4271/9781468606003

ISBN-Print 978-1-4686-0599-0
ISBN-PDF 978-1-4686-0600-3
ISBN-epub 978-1-4686-0601-0

To purchase bulk quantities, please contact: SAE Customer Service

E-mail: CustomerService@sae.org
Phone: 877-606-7323 (inside USA and Canada)
724-776-4970 (outside USA)
Fax: 724-776-0790

Visit the SAE International Bookstore at books.sae.org

Publisher
Sherry Dickinson Nigam

Product Manager
Amanda Zeidan

Production and Manufacturing Associate
Michelle Silberman

Contents

Preface

In the year 2000, I considered myself extremely fortunate to break into the "Configurations and Engineering Analysis" group within Boeing. C&EA, as it was also known, was the airplane design group where innovative ideas take root within Boeing Commercial Airplanes. I was assigned to the Boeing 777-300 Extended Range project, and it soon became apparent that there was a lot to learn about in every aspect of the airplane design process. At the same time, as the 777-300ER was in development, Boeing was also doing extensive studies on the "Middle of the Market" or MOM airplane. The new airplane design group was in competition with 767 and 777 re-engine, re-wing studies to fill the market design space, as so often happens when a new market need is identified.

Contrary to popular belief, it is not a foregone conclusion that an all-new airplane is always the right solution for an emergent market need. A sizable portion of an aircraft's performance typically comes from advances in the wing design and propulsion system. Hence, often, it is cost-effective to stick with an existing fuselage design, which is then mated to a new wing and engine. As part of the Boeing 777 aircraft design group, I got involved in the 230- to 290-seat, re-wing and re-engine of the 777 MOM family.

Wing design was a fascinating new world for me, and it was at this time that I got exposed to the bizarre reality that on commercial transports, landing gears drive the wing design as much as aerodynamics. The complexity of the science of wing design immediately drew me in, and I was hooked. Every airplane designer I spoke to at Boeing and later at Airbus had a different approach to the design problem. I soon realized that there was as much art as science to the design of wings, and the key to it all was to be flexible and understand how the merits of a wing design could influence and be influenced by various aspects of the airplanes components, be it the fuselage, empennage, landing gear, or the propulsion system. In short, there is no "silver bullet" to a design problem, and what works well for one design may not be optimum for a different aircraft design solution.

My inspiration for this book is born out of my 23+ years as a commercial aircraft designer, and the desire to share a different more intuitive approach to jet transport design. I have combined my work experience with the expertise of several authors who have written some superb books on this subject, to develop new and novel methods for my book. This book deals with how to think about commercial aircraft designs and also how to layout an airplane. For obvious reasons, I have steered away from proprietary formulations used by the major manufacturers but, at the same time, endeavored to give the reader a glimpse into the philosophy and approach to jet transport design as practiced in industry.

This book is for all you aircraft enthusiasts. Whether you are a pilot, an aviation executive, an airliner enthusiast, or an aerospace professional, you should enjoy this book. While I have failed miserably in keeping with my original goal of making this an equation-free read, I believe I have succeeded in exposing you the reader to how commercial aircraft are designed and why they look the way they do—I sincerely hope you enjoy the read.

Foreword

Eleven years ago, while I was teaching at IIT-Bombay, Zarir had been working on the initial parts of "Jet Sense." I had just left Boeing and Zarir offered me the empirical equations he had developed to use in my class notes. Since then, people have been waiting for this book and not just because of the data and empirical equations. Zarir's book offers a complete overview of preliminary airplane design from within the trenches of a large commercial aircraft manufacturing company.

I think there are several things that are special in "Jet Sense." The first is that it starts by focusing on the market. When teaching airplane design, I would spend the first two lectures in my design classes (at several universities) discussing the market. If the airplane does not fit the market the airlines want, it will never sell. It is important for commercial airplane designers and students to understand the compromises needed to design and build a sellable airplane. Was the Airbus A380 a failure? It is a marvelous airplane with state-of-the-art technology. But it was not nearly the market success of Boeing's 787. "Jet Sense" emphasizes this example before continuing to the meat of the book.

Another special feature of Zarir's book is that it emphasizes geometry and integration in a way that I have not seen in other texts. Integrating the landing gear into the wing may be the most difficult aspect of designing a commercial airplane. I would tell students this at the beginning of the course, but most would only discover that reality when their design would not integrate properly. Integrating the wing with flaps, slats, spars, etc. has not really been complete in other texts, which focus more on the analysis. I feel Zarir's book helps illustrate that challenge and gives guidance. It also lays out certification requirements when needed.

I also think that Zarir's book follows a unique approach to initial sizing, which is suited for commercial airplane design. This is more likely to give better and quicker solutions. Traditional approaches emphasize maximum aerodynamic efficiency, minimum fuel consumption, and lightest weight. Seldom are the trade-offs tied to what the airplane actually is supposed to do. Zarir differentiates between single-aisle work horses, like the Boeing 737 and Airbus A320, verses long-haul twin-aisle aircraft, like the Boeing 787 and the Airbus A350.

For about five years, I carpooled with Zarir from North Seattle to Boeing in Everett, Washington. Many of those trips were the most fun and stimulating times in my career. We talked about what we would do if we were "The Boss." We discussed changes we would make to Boeing's biggest twin, the 777, in a new derivative. It pleased me to see Boeing taking that direction, even though the final outcome was very different to what Zarir and I had bantered about. In that daily commute, Zarir helped me gain confidence and trust in our collective knowledge of commercial airplane design. Zarir really understands the trade-offs needed to find a sweet spot for a market. His collection of data, taken over years from various public sources, would get him to solutions that I could trust. Those data went into the equations used in this book.

"Jet Sense" focuses on commercial aircraft. It is not an introductory aircraft design book covering all types of aircraft. But for commercial aircraft designers, this should be on every designer's desk.

—**Scott Eberhardt Ph.D.**
Aerospace Consultant and Author of *Understanding Flight.*

Acknowledgements

This book reflects my years of experience in the commercial aerospace industry both in the United States and in Europe where I have been fortunate enough to work in the capacity of an aircraft designer. It has been motivated by the absence of literature that focuses on configuration design and layout as practiced in the commercial aircraft industry.

This book is dedicated to my family beginning with my wife Tejal who has been my confidant and friend through the trials and tribulations that are now the norm of the aerospace industry. My children Zayn and Zara, who have brought joy to my life and who keep me working towards a brighter and cleaner future for generations yet to come. My parents Yasmin and Dinshaw, who have encouraged my interest in aviation since a very young age and who have supported me by providing me with an education in the United States. My sister Dilshad, whose moral support through my years in academia has kept me on task when a future in the aerospace industry looked bleak in terms of a career prospect.

It has also been my privilege to have worked with some of the brightest minds in the industry; without whose guidance and experience this book would not be possible. I thank, Mike Sinnett, Terry Beezhold, Kourosh Hadi, Brian Nield, Larry Schneider, Lars Anderson, David Loffing, Robb Gregg, Matt Orr, Brian Hermesmeyer, James Knapp, Bob Wild, Michael Drake, Michael Burtle, John Dixon, Ben Almojuela, Rhodri Thomas, James Freitas, Mithra Sankrithi, Gerhard Seidel, Edward Petkus, Christopher Musoke, Timothy Purcell, Robert Hoffenberg, Thomas Ahn, Siddharth Banerjee, Hans Fischer, Yvon Vigneron, Robert Davies, David White, Andy Ouellette, Robert Nelson, Wendell Choy, Warren Steyaert, Daniel Wajerski, Gnanulan Canagaratnan, Scott Eberhardt, Aspi Eranee, Giles Khan, Phiroze Khan and Praveen Kamath for their time and their patience in educating me through my journey as an aviation enthusiast and an aerospace engineer. Last but not the least I would like to thank Paul Wojciechowski for his mentorship and in giving me the curiosity and quest for knowledge, that set me on a journey to seek out innovative and simple solutions to complex problems.

Pusteflower9024/Shuttersto

Airport Departure Board

1

Markets and Requirements

We were gathered mid-field on the south side of runway 27 to witness the final flight of the Concorde touchdown at its birthplace in Filton, England. The afternoon sun framed the Rolls Royce factory and the surrounding neighborhood in a perfect backdrop to capture what was certain to be a defining moment in aviation history. Rolls Royce and the British Aircraft Corporation were instrumental in the birth of Concorde, and it was only fitting that this marvel of commercial aircraft design was coming home to this location on its final flight.

FIGURE 1.1 Concorde returning to Filton, UK, on its final flight.

Matt Cardy/Alamy Stock Photo.

It occurred to me that as the chapter on commercial supersonic flight was coming to an end, commercial transonic transports were on the cusp of achieving new technological heights. In the US, the Boeing Company had boldly embarked on the development of the all composite 7E7 (the 787 as we now know it today), while in response, Airbus was furiously designing the A30X (or what would eventually become the A350) to close the gap in technology. Commercial aircraft design was on the eve of a paradigm shift, and it was thrilling to be a part of this moment in aviation history. It was 2003 and I was an advanced design engineer at the Airbus UK, Future Projects Office. The new millennium held great promise for the commercial aircraft industry, and the game between the two aerospace giants was on.

While the Airbus A380 was being developed and built, it was the last of a generation of airliners based on existing propulsion, material science, and systems technologies. These technology choices ultimately led to the demise of the A380. Boeing had placed a different bet. The 787 incorporated advanced composite, propulsion, aerodynamics, and systems technologies to make the long-range, medium-capacity airliner a possibility. Boeing bet, correctly, that point-to-point travel would be what the traveling public would prefer and that new routes would ultimately help decongest hub airports such as London Heathrow. Airbus, on the other hand, opted for an increase in aircraft size to drive per seat improvements in operating cost and address the congestion issue differently.

Predicting changing and emerging market trends are an integral part of getting the requirements for a new jetliner right. While in the past, forecasting the demand for jets 20 years in advance used to be fairly straight forward, the introduction of new low-cost carriers (LCCs) combined with soaring fuel prices has changed the operating landscape. A small miscalculation in requirements can be disastrous, as experienced by the A380 program, which ultimately had a program life of just 14 years.

Airbus has clearly internalized the hard lessons of the A380 program. The advent of the A350-XWB is proof that the airframer is coming of age and has realized the need for a successful long-range, wide-body twin engine jet to compete effectively in the market place. The European aerospace industry has successfully evolved into a force to be reckoned with. Both Airbus and Boeing understand the need for good payload-range and environmental performance (noise and emissions). The current market outlooks (CMOs) of both manufacturers are based on extensive research and are the benchmark used by the aviation industry and the financial institutions that support it. While Airbus and Boeing market outlooks tend to have a slight bent toward promoting their individual product lines, all in all, they are fairly consistent in how they view the airliner needs of the future.

Looking at the Boeing 2016 CMO, we can begin to dissect it to gain more insights into where future market opportunities could exist. Combining the CMO data with historic replacement cycles and engine technology maturation timelines, one can begin to formulate what a future aircraft requirement may look like. Using this approach, coupled with an average growth in world air traffic of 5% can help us predict the replacement market fairly reasonably. Predicting emerging trends can be a bit more tricky, but not impossible. The attached chart provides some insights into the future market need and also the existing gaps in the seat-mile design space.

Creating additional subcategories of the Boeing CMO data is shown in Figure 1.2. The figure gives us some basic information of growth and replacement units in each category and the number of new units needed in the 2015 through 2035 time frame. Even though the individual growth and retirement rates per year vary with the category of market, we can derive an average rate for growth and replacement for the market in its entirety.

FIGURE 1.2 New airplanes required between 2015 and 2035.

New Airplanes Required 2015–2035

Boeing 2016 CMO—Adjusted

Created new category

Category	Units in 2015	Units required for growth 2015–2035	Units to be retired 2015–2035	Total number of new aircraft needed 2015–2023
Ultra Large Twin Aisle	340	360	170	530
Large Twin Aisle	2040	1650	1820	3470
Medium Twin Aisle	2052	2623	1311	3934
Small Twin Aisle	608	777	389	1166
Single Aisle	14870	17,410	10,730	28,140

- Ultra large twin aisle 420 plus seats long range dual class arrangement [B747-8, A380-800]
- Large twin aisle 321–420 seats long range dual class arrangement [A350-1000, B777-300ER, B777-9,B747-400, B787-10]
- Medium twin aisle 245–320 seats long range dual class arrangement [B767-400ER, B787-8,-9, B777-200, A330, A350-900]
- Small twin aisle 200 seats long range dual class arrangement [A310/767-300ER]
- Single aisle 90+ to 230 seats short range dual class arrangement [ERJ 190 thru A321 NEO, B757]

Average commercial aircraft fleet growth/year over 20-year period = 5.6%
Average commercial aircraft fleet retirements/year over 20-year period = 3.4%

Further refining and plotting the categories by seats and range give more insights on the average number of seats and how far they fly. Moving the 787-8 and the A330-200 into the same category as the 767 family creates a more refined payload-range picture in this category of airplanes.

Similarly, creating a separate category for the 787-9, A350-900, B777-200ER, and A330-300/A330-900 NEO helps further refine this payload-range category.

Lastly, the 747-400 fits better with the 787-10, 777-300ER, and A350-1000 and the 777X family. Recognizing that the 777-9 is a bit larger than the 777-300ER and similar in size to the 747-400. The new twin engine airplanes bring improved economics into a category once exclusively dominated by four engine jets.

FIGURE 1.3 Airplane type and market.

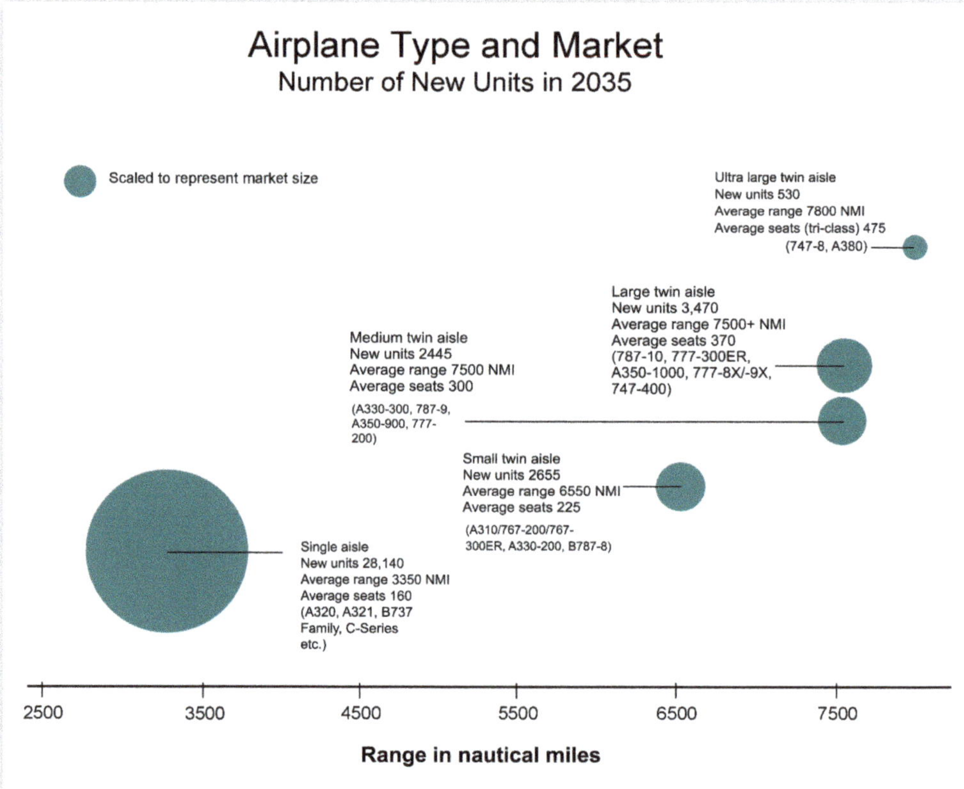

Assessing this plot, we can draw the following conclusions:

The single-aisle market has the maximum number of units to be gained in the next 20 years, and on average, these aircraft have a seat count of ~160 seats (typical short range seating arrangement) and a range of ~3350 NMI.

Several new entrants such as the Chinese C919, the Russian MC-21, and the Canadian C-Series further confirm that this is a lucrative area of growth in the commercial airplane market. In this seat count segment, there are enough new and replacement units for new entrants to justify the financial risks of undertaking a new aircraft development program.

Boeing and Airbus themselves have developed upgraded versions of their 737 (The MAX) and the A320 (NEO or New Engine Offering) to capitalize on this design space

The next big jump in the number of new units forecast in 2035 is in the Large Twin Aisle space in the 370-seat category (typical long range dual-class seat count). These aircraft have an average range of 7500 NMI in the dual-class configuration. Typically, at these higher seat counts (greater than 350 seats dual class), these aircraft are also configured in a tri-class (first, business, economy) layout, to further optimize the revenue for an airline. A tri-class configuration reduces the dual-class seat count by ~15%. This reduction in seats can translate to a range increase of 300 to 500 NMI. An anomaly in this category would be the ultra-long-range 777-8X, which has a range in excess of 8800 NMI in a dual-class configuration. This category of "ultra-long-range" aircraft offer a unique capability but have a very limited market, historically totaling less than a 100 units.

The 300-seat category also has a substantial number of units forecasted by 2035. This category has seen the advent of the B787-9, the A350-900, and the A330-900NEO/A330-300. This category has become the mainstay for a lot of the world's airlines. Their larger stretch versions in the 370-seat category serve predominantly as flagships.

There is also a reasonable market for twin-aisle aircraft in the 225-seat category (dual-class long-range seat layout). These aircraft have a range of 6550 NMI. The B767-300ER range is on the lower end of that category, while the B787-8 and A330-200 are in the upper end of the category.

The smallest number of new units required in 2035 is in the ultra-large twin-aisle category, which is currently dominated by the A380-800 and the B747-8i. This forecast is consistent with previous forecasts, which have predicted a decline in this category relative to the large twin-aisle category. It appears that pundits on both sides of the debate for ultra-large airplanes now agree that there is a need for aircraft in this category, although whether they should be as large as an A380 is debatable. Perhaps the A380 was ahead of its time, and as airport congestion continues to increase past the 2030s, one may see a revival of the ultra-large aircraft. Technological evolution could also see this category of aircraft become more economical to operate and see a smaller than 500-seat aircraft materialize in this design space. However, if the trend of decreasing aircraft size in favor of frequency continues, it is entirely possible that the ultra large twin-aisle category of aircraft could go away and be replaced by aircraft such as the Boeing 777-9 and the A350-1000. Both these aircraft have large enough wings, which when mated with future engine improvements could be stretched even further with an increase in payload and range capability.

Examining these categories in terms of the payload-range of the individual products currently flying or in the process of entering service paints an interesting picture. Roughly 85% of the world's long-range routes can be served by a range of 6100 NMI. The slope of routes covered per hundred nautical miles is fairly steep up to a range of 7300 NMI. The Boeing 787-10 which has a range of 6350 NMI covers approximately 88% of the world's routes, although at the time of this writing, Boeing did announce an increased take-off weight version that would likely push it to 6700 NMI. A range of 7250 NMI makes up 98% of the world's long-range routes. This is an interesting data point as for the longest time the B747-400 dominated the long-range market, with a range of 7200 NMI. For its time, the 747-400 had unmatched capacity and excellent operating economics; however, it is widely accepted that the main reason for its success was because of its unmatched range capability. The 777-300ER had equivalent range and twin-engine operating economics, and even though it did not match the 747-400 in size, it eventually was responsible for the demise of the iconic Boeing jet.

FIGURE 1.4 Twin-aisle products and typical routes flown.

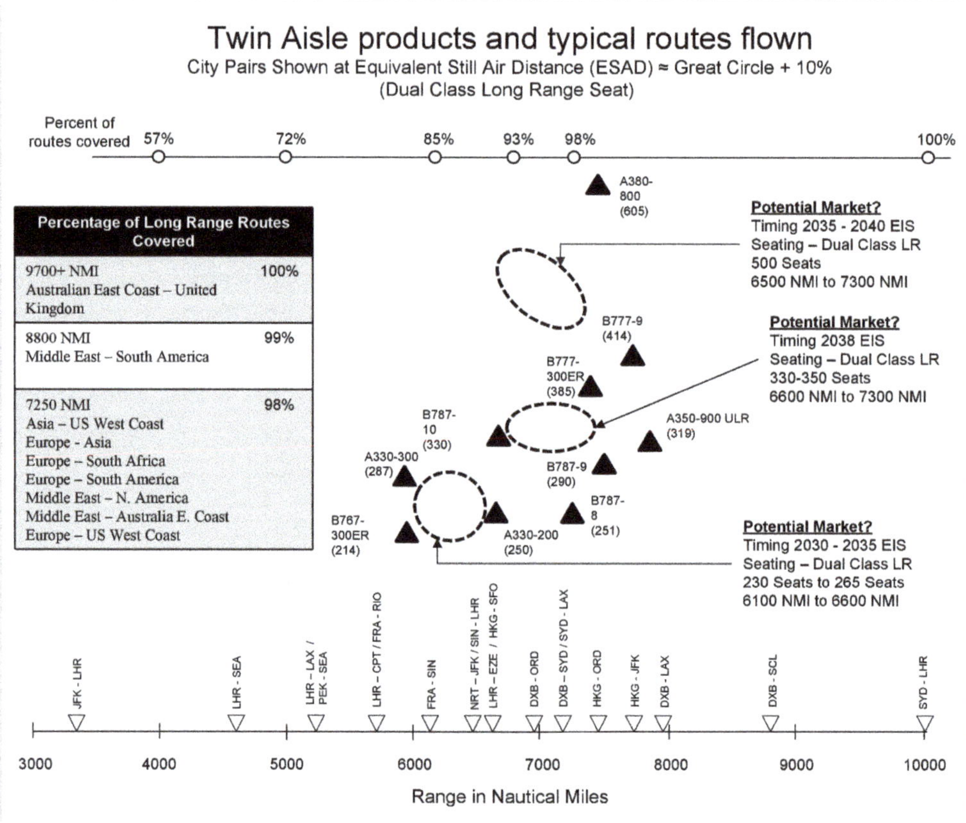

As technology has enabled smaller twin aisles to be more economical on long-range routes, airlines have used airplanes such as the B767-300ER, A330-200 (early 1990s), and the Boeing 787-8/-9 (mid-2010 onward) to open up new longer range routes, which could then be swapped for larger jets, once the market was mature and passenger traffic levels stabilized. As new routes from the Middle East to North America have evolved, so has the mission capability of the aircraft serving them. The Boeing 777-9 being the pinnacle of what a large twin-engine jet is capable of in terms of payload-range and operating economics.

This strategy while successful has, rather ironically, left a gap for an efficient mid-range aircraft. The B767-300ER is a 1980s vintage aircraft, and the A330-200 and Boeing 787-8 carry a large wing and engine for increased payload-range capability and also growth. Both these aircraft, while initially envisioned to replace the B767-300ER, carry a large wing area in the ballpark of 3850 ft^2 and engines capable of 70,000+ lb of thrust. While these airplanes are economical to operate, their capability demands a price premium—a necessity given that the increased capability is enabled by a larger wing and engine, which are also expensive to build.

Looking at the payload-range charts, one could conclude that an airplane having a range of 6600 NMI, seating 230 passengers (in a dual class long range) might be a reasonable space to start an aircraft family. This aircraft could have growth to 265 seats (dual class long range) and fly 6100+ NMI, essentially covering the market segment of the B767-300ER/A330-200 and the A330-300 and up to 90% of the world's routes.

More importantly, this aircraft would carry a smaller wing and engine and, hence, would be cheaper to manufacture. The timing of this aircraft would likely have to coincide with when the next step function improvement in engine technology is made available.

Figure 1.5 shows when new engines have become available historically and when the next improvement could be expected for a given a thrust class.

FIGURE 1.5 Engine thrust class and availability.

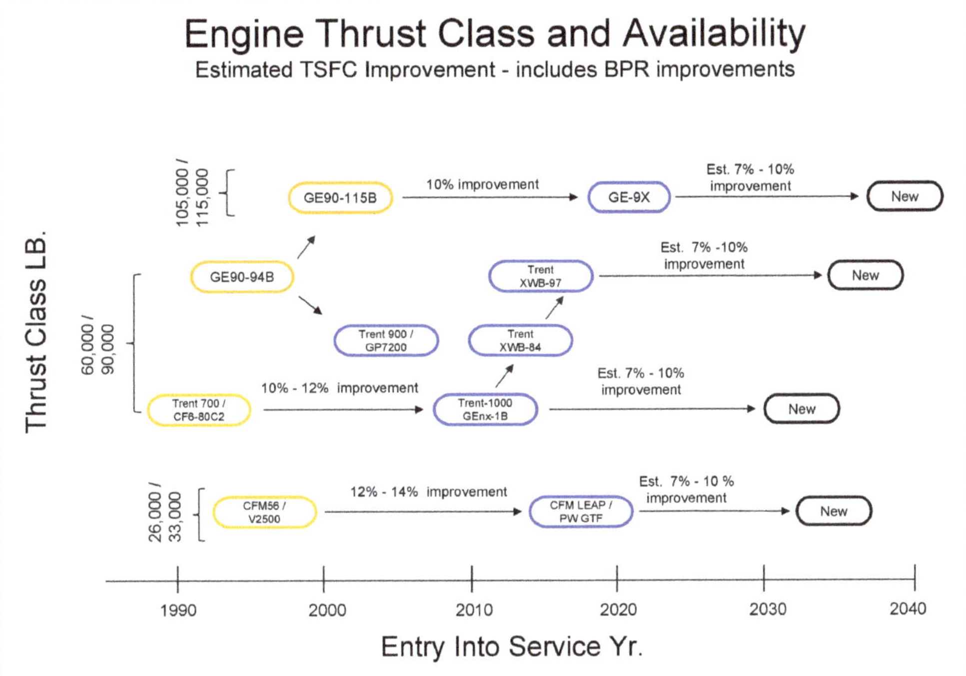

The underlying assumption is that engine technology improves at a rate of 0.5% a year. Application of this trend should be used with some degree of caution. What the trend does not imply is that if you wait, every two years you may get a 1% improvement in specific fuel consumption. Engines like airplanes require a substantial investment in technologies such as material science, aerodynamic modeling, testing, and systems. Engine manufacturers usually come up with a new engine every 15 to 20 years, allowing for a 7 to 10% improvement in engine-specific fuel consumption. Embedded in this improvement is an increase in the engine bypass ratio. The bypass ratio typically increases because of a combination of technologies that reduce the diameter of the engine core and/or enable an increase in the fan diameter. This trend chart would indicate that the next improvement in engine technology in the 60,000-lb thrust class would be sometime in the 2030 through 2032 time frame. If one were to be conservative, one could then say that a medium-range, efficient transport would be viable in the 2030 time frame—give or take a couple of years.

This is an example of how one might approach looking at a market design space and develop requirements. As mentioned earlier, requirements can play a crucial role in shaping an aircraft family and its success over a 20-year time period.

Another important consideration from a manufacturer's perspective is the profit potential of a family of designs. Early on in the aircraft design process, the most common metric used to understand competing designs is operational cost. While one of the biggest levers in the operating cost of an airplane is fuel, fleet decisions are made on several factors including but not limited to availability, airplane price, industrial offsets, global airline partnership agreements, and national interest. Fuel burn plays into the operational cost, while payload and range impact the revenue potential of the product. Both operational costs and revenue potential impact the price of the aircraft, which has a direct effect on the initial cost of ownership. More often than not, price is a lever that is used to nullify an operating cost advantage related to fuel and other value drivers. For a manufacturer to take advantage of this, the airplane design department should seek metrics that not only take price into account during the initial sizing phase but also the recurring cost to build the airframe.

Aircraft pricing models are closely held by different manufacturers; however, with a little bit of judgement and information on existing list prices, one can develop a time-sensitive pricing model. Historically, data shows that a price escalation of 1.1% a year is a reasonable assumption.

Airframe costs, on the other hand, correlate well with airframe weight. Airframe weight is defined as aircraft empty weight minus seats and dry engine (engine without nacelle). Airframe costs can vary from 250 to 300 $/lb [11, 66, 70, 71]. Several factors that impact the range of costs include but are not limited to the following:

- Labor rates
- Automation
- Raw material costs
- Transportation costs or logistics
- Geopolitical location
- Quality control practices
- Design for build practices
- Supplier relations and contracts
- Government offsets to do business in a foreign country
- Other

Typically, on a dollars per pound basis, composite airplanes tend to be at the higher end of the scale for several reasons which can include the following:

- Raw material costs, which are higher than aircraft-grade aluminum metals.
- Carbon fiber manufacturing technologies. In the commercial aircraft industry, CFRP manufacturing methods are conservative and incorporate expensive traditional processes.
- The net effect is that the airframe is lighter and the dollar per pound metric can also inflate the number (weight being in the denominator).

For early design purposes, it is acceptable to assume that a composite airplane will cost up to 10% more than a traditional aluminum airplane on a dollars per pound metric. Having said that, there is no substitute for common sense, and if detailed cost numbers are available, the most current data should be used instead. Much like the pricing model, costs can also be estimated to escalate at 1.1% a year.

When considering operational costs and timing, especially at a future entry into service date, an assumption has to be made on fuel price. While this is the equivalent of looking into a crystal ball, the chart in Figure 1.6 can be used to pick a reasonable number.

FIGURE 1.6 Jet fuel prices over the past 32 years.

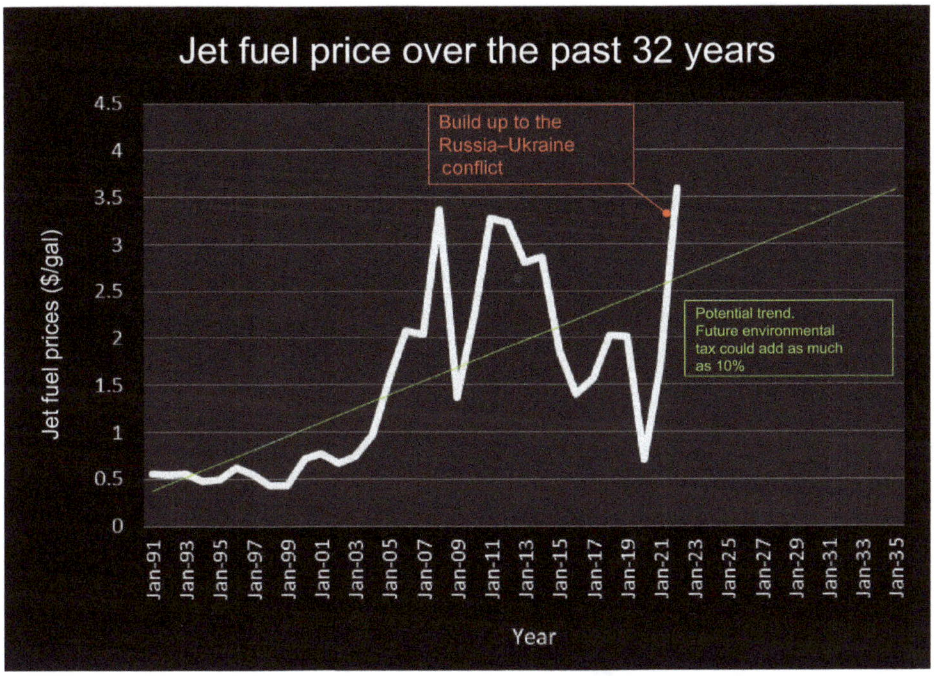

Taking the 230-seat aircraft example, one can further build up a set of airplane-level requirements for initial configuration studies. Requirements for this category of transport may be given as follows:

- Entry into Service 2030.
- Family: Two-member family.
- Seats: Seat count separated by 15% plus, starting at 230 seats in dual class (Stretch at 15% = 265 seats).
- Range: 6600 NMI for 230 seats and greater than 6100 NMI for the stretched airplane.
- Engines: Available technology in 2030 time frame.
- Cruise Speed: Mach 0.84.
- Wing Span: Code E < 213 ft. Objective for Code D < 171 ft (A310 and B767 Gates).
- Fuel burn per seat economics on 3000 NMI mission: 15% better than best in class.
- Take-Off Field Length: Less than 9300 ft in ISA+15 conditions for the stretch airplane.
- Approach Speed: Less than or equal to 145 knots (stretch airplane), 140 knots (initial model).

Deep conversation in an advanced design office
If I don't know what your airplane looks like, how can I start sizing it??
If I don't know what size it should be then I can't draw it!
Future Projects Office

© Zarir D. Pa

2

Aircraft Sizing

Aircraft sizing, as the term implies, involves sizing the major components of the aircraft such as the fuselage, wings, engines, and empennage to meet a specific set of requirements. These requirements may initially be presented to the designer as airplane requirements and objectives (AR&O) as it is known in the industry. The sizing or design point selected by a design team locks in a significant portion of the weights, propulsion, systems, and aerodynamic characteristics of the aircraft. Hence, its selection must be given a lot of thought in terms of the proposed initial mission and the final family plan for the design. The multitudes of sizing methods available in the aerospace industry constitute extremely sophisticated computational algorithms which allow the design engineer to fix the payload and range for an aircraft family, and then iterate to the wing area that achieves the best compromise. It is not a coincidence that during the writing of this book, both Airbus in Europe and Boeing in the US are pursuing their own versions of integrated sizing tools.

The guiding equation that governs aircraft sizing is the Breguet range equation. While this equation is well known within the aeronautics community, it is worthwhile to spend some time in understanding this equation and how it relates to the aircraft geometry and the sizing process.

The Breguet range equation for a jet is given by the following equation:

$$R = a \times M / TSFC \times L / D_{cruise} \times \ln\left[Wt_{initial} / Wt_{final}\right] \tag{2.1}$$

where
R is the range in nautical miles (NMI)
a is the speed of sound at cruise altitude (knots)
M is the Mach number at cruise altitude
TSFC is the specific fuel consumption (lb/hr/lb)
L/D_{cruise} is the lift-to-drag ratio in the cruise condition

Here,

$$L = \text{Lift}\,(\text{lb}) = \text{Weight in steady}-\text{state flight} = W = C_L \times q \times S_{ref} \tag{2.2}$$

where
C_L is the lift coefficient
q is the dynamic pressure (lb/ft^2)

$$q = \left[\tfrac{1}{2} \cdot \rho \cdot V^2\right] \tag{2.3}$$

where
ρ is the density in slug (ft^3)
V is the velocity (ft/sec)
S_{ref} is the reference wing area (ft^2)

$$D = \text{Drag}\,(\text{lb}) = \text{Thrust in steady}-\text{state cruise flight} = C_D \cdot q \cdot S \tag{2.4}$$

where
C_D is the drag coefficient
$Wt_{initial}$ is the initial weight at start of cruise (lb) = MTOW for all practical purposes, where MTOW = maximum take-off weight for the aircraft
Wt_{final} is the final weight at the end of cruise (lb) = OEW + payload + reserve fuel load, where OEW = operating empty weight = empty weight + crew + trapped fuels and oils

Each term of the Breguet range equation is related to the aircraft geometry and its aerodynamic and structural characteristics. One can influence the design range and the fuel burn of the aircraft by understanding and then trading these variables to achieve a balanced design.

Let us examine each of these variables in more detail.

Speed of Sound

The speed of sound, a, is dependent on the altitude selected for cruise. For a standard day, the speed of sound is a function of air temperature, and as such, it decreases linearly up to an altitude of 36,089 ft, which is also known as the tropopause (where the troposphere ends). Beyond this altitude, the temperature remains constant up to 80,000 ft. This is the stratosphere. Hence, beyond 36,089 ft, the speed of sound is constant for all practical purposes at the altitudes that a jet transport flies. What this also means is that at a constant

Mach number, the true airspeed (TAS) of the aircraft increases with altitude up to the tropopause, after which the TAS remains the same.

Cruise Speed

The cruise Mach number, M, can be chosen in several ways. Usually this number is derived from a marketing exercise, or it can simply be chosen based on the need to match or improve upon the competitors' cruise performance. Most modern jets are now designed for cruise Mach numbers that can range between 0.6 for short-range regional jets, and 0.92 for long-range flights. One may choose to fly faster than their competitor, but there is usually a price to pay in fuel burn and in aerodynamic efficiency. Even with all the advancements made in the aerospace industry, manufacturers on both sides of the Atlantic Ocean often struggle to put a dollar value on speed. The question often is does one fly faster and charge passengers a premium fare or do you design for a slightly lower speed and strive for an improved economic efficiency?

Boeing recently revisited this question and attempted to answer it when it decided to boldly take on the sonic cruiser program. The sonic cruiser was reputed to have a 0.95 Mach cruise speed which promised to cut travel time on long-range flights, especially across the vast Pacific. Boeing did ultimately manage to conquer the various technical challenges associated with the chosen design point so that the fuel burn of the aircraft was comparable to existing transonic transports. However, the airlines opted instead for the 787, which was being developed on a parallel path and which promised fuel efficiencies for mid-sized jets unrivaled in the industry. The 787 will cruise at 0.85 Mach when it enters into airline service in 2008. This choice of cruise Mach will let the 787 operate on the same routes and within the same block times as the Airbus A380, the Boeing 747, and 777, which operate at cruise Mach numbers in the range of 0.84 to 0.86.

The maximum operating Mach number, M_{mo}, is the maximum cruise Mach number for the aircraft. The M_{mo} offers the operator some mission flexibility such as to make up for a delay on the ground or to make up lost time for excessively high headwinds. The maximum operating Mach number can be set by a number of different factors that have to be taken into consideration. These factors typically include but are not limited to Mach buffet, off-design stability and control criteria (should the aircraft lose one or more hydraulic or electrical flight control systems), design dive speeds, and the maximum thrust available.

Typically, the M_{mo} is set by a performance requirement, and more often, it is governed by the design philosophy of the aircraft manufacturer. Airbus, for instance, sets M_{mo} as $M_{cruise} + 0.04$, while Boeing tends to provide the operators a bit more flexibility by setting M_{mo} as $M_{cruise} + 0.05$. M_{cruise} and M_{mo} also set another very important Mach number, which is the dive Mach number, M_D. M_D is often 0.07 Mach above M_{mo}, but it can be reduced to $M_{mo} + 0.05$ if the manufacturer can prove that the aircraft can be recovered from a 7.5° nose down dive within 20 sec before M_D can be exceeded. The 747 and earlier jet transports were designed and certified in a time when the FARs defined $M_D = M_{mo} + 0.05$. There have been several incidents such as the China Airlines 747 which encountered an upset at cruising altitude. It is suspected the airplane may have broken the sound barrier as it dived toward the ocean. The airplane recovered at about 9000 ft, and the structure while still intact showed signs of deformation and damage. The 747 landed safely—A testament to the structural integrity and capability of this particular aircraft design. Table 2.1 lists the various design Mach numbers of some current generation transonic jet transports.

TABLE 2.1 Design Mach numbers of some current jet transports.

Aircraft	M_{cruise}	M_{mo}	$M_D = M_{mo} + 0.07$
A340-300	0.82	0.86	0.93
A350-900/-1000	0.85	0.89	0.96
A380-800	0.85	0.89	0.96
B777-300ER/777X	0.84	0.89	0.96
B747-400	0.855	0.92	0.97 (M_{mo} + 0.05)

Thrust-Specific Fuel Consumption

Thrust-specific fuel consumption, TSFC, is a measure of the fuel efficiency of the jet engine per pound of thrust it produces. It is defined as the ratio of fuel burned per hour to the pounds of thrust produced by the engine. It has units of lb/hr/lb = 1/hr.

TSFC is a function of various engine parameters, the most prominent of all being the bypass ratio. Most modern jet engines are usually composed of two major sections which break up the oncoming air: the core of the engine, which is composed of the compressor, combustion chamber, turbine, and exhaust, and the bypass section, which is simply a contoured duct that surrounds the core and which routes the air from the fan straight through the nozzle after accelerating it. Hence, the bypass ratio is defined as the ratio of the "bypassed" air to the air that passes through the engine core. Bypass ratios for modern jet engines are on the order of 5:1 and can range up to 10:1 with the promise of even higher bypass ratios (12:1) in the future. Typically, the higher the bypass ratio, the lower the TSFC for the same pound of thrust produced. Hence, most modern jet engines such as the Pratt & Whitney Geared Turbofan (GTF), Rolls Royce Trent series, and General Electric turbofans have large diameter fans.

TSFC can range from 0.64 for older generation small engines in the 13,000 lb thrust class found on the in-service regional jets to ~0.55 seen in bigger jets like the A380 and the 777. Modern jet engines being used on the Boeing 787 and the Airbus A350 deliver even higher efficiencies. The GE9X engine that powers the 777X family of airplanes has TSFCs that have a further improvement over the latest generation of engines on the A350 and the Boeing 787.

Lift-to-Drag Ratio

The lift-to-drag (L/D) ratio is a measure of the aerodynamic efficiency of the aircraft. All else being equal, the higher the lift-to-drag ratio of the jet, the greater is its range performance. Conversely, for a given range, the better the lift-to-drag ratio, the better the fuel burn. It can also be written as follows:

$$L/D = C_L / C_D \tag{2.5}$$

At this point, it is appropriate to introduce the drag polar, which, in its simplest form, can be defined as follows:

$$C_D = C_{D0} + C_{DI} + C_{Dwave} \tag{2.6}$$

where
C_{D0} is the profile drag coefficient {function of aircraft geometry and shape}
C_{DI} is the induced drag coefficient {lift-dependent drag coefficient that is a function of span loading or weight/wingspan}
C_{Dwave} is the compressibility drag {function of wing sweep, body fineness ratio, and wing thickness ratio}

As we begin the sizing process, we will examine each of these coefficients in more detail.

Typical values of C_L for jet transport can range from 0.45 for a very large long-range transport to 0.55 for relatively slower and smaller short-range jet transports. Typical values of C_D for a jet transport are on the order of 0.031 for a narrow body and 0.0240 for a wide body. This equates to 310 drag counts and 240 drag counts, respectively, where 1 drag count is defined as 0.0001.

Profile Drag

Profile drag of the airplane is the drag due to the aircraft geometry or profile. This includes drag due to skin friction and pressure drag.

Here,

C_{D0} is the profile drag coefficient
$= C_{D\text{-friction}} + C_{D\text{-pressure}}$
$C_{D\text{-friction}}$ is the friction drag related to the exposed surface area of the airplane
$C_{D\text{-pressure}}$ is the pressure drag related to the shape of the aircraft component such as thickness and curvature, which is also commonly known as form drag

$$C_{D0} = \Sigma C_f \cdot A_{wetcomp} / S_{ref} \times FF$$

where
$A_{wetcomp}$ is the exposed wetted area of the aircraft component (wing, fuselage, tail, etc.)
S_{ref} is the wing area
C_f is the skin friction coefficient
FF is the geometry-related pressure or form drag factor

$$\text{For wings} \quad FF = \left[1 + 2.7(t/c) + 100(t/c)^4\right] \quad \text{for wings and tails} \tag{2.7}$$

$$\text{For fuselages} \quad FF = \left[1 + 1.5/(L_f/D_f)^{1.5} + 7/(L_f/D_f)^{3.0}\right] \tag{2.8}$$

Altitude Effects Related to Profile Drag

Most performance calculations are run at 35,000 ft as this is the altitude at which transonic jets show a good match between efficient engine fuel burn characteristics and aerodynamic efficiency. The L/D ratio decreases with increasing altitude due to the Reynolds number effect. For a given engine, the fuel consumption typically improves with altitude up to the tropopause (36,089 ft), after which the fuel burn tends to decrease with altitude.

Why is this important? The Reynolds number, named after Osborne Reynolds (1824–1912, British National), is a nondimensional number which is defined as the ratio between the inertial forces to the viscous forces.

$$Re = \rho \cdot V \cdot L / \mu \quad (2.9)$$

where
Re is the Reynolds number
ρ is the air density in slugs (ft^3)
V is the local velocity (ft/sec)
L is the characteristic length of the component being analyzed (ft)
μ is the dynamic viscosity in slugs (sec)

The Reynolds number allows aerodynamicists to corelate the operational conditions of different aircraft to each other in a nondimensional manner for analysis purposes. Hence, the aerodynamic characteristics of a wind tunnel test model can be correlated to an aircraft in real flight if they are both operating at a similar Reynolds number.

For $Re < 5 \times 10^5$, the flow is said to be mostly laminar. Laminar flow is usually associated with a lower drag level. Above this number, the flow transitions to a turbulent flow field usually associated with a higher drag level. Most transonic jets operate in the Reynolds number range on the order of 3×10^6 or greater and are hence in a fully turbulent flow field. As can be seen from Equation (2.7), the Reynolds number is directly proportional to the air density, which decreases with altitude. The skin friction coefficient, C_f, is related to the Reynolds number by the following equation:

$$C_f = 0.455 / \left\{ \left(\log_{10} Re\right)^{2.58} \left(1 + 0.144M^2\right)^{0.58} \right\} \quad (2.10)$$

In Equation (1.8), we see that as the Reynolds number decreases, the skin friction coefficient increases, but as the Mach number increases, the skin friction coefficient decreases. This is an important effect. Equation (1.8) says that for a given aircraft flying at a given Mach number, as we climb higher, the maximum lift-to-drag ratio will decrease. In other words, the profile drag, which is a function of the skin friction coefficient, increases with altitude due to the Reynolds number affect.

The aforementioned relationships are evidence that the choice of cruise altitude for the design point can have a significant bearing on the size of the wing and the engine thrust required for efficient cruise flight.

The lift-to-drag ratio is also a function of C_L, as shown in Equation (2.5). Rewriting the lift equation, Equation (2.2), we can obtain a correlation between C_L, the wing loading, Mach number, and the cruise altitude as follows:

$$C_L = (W / S) / \left\{ 1481.3\delta \cdot M^2 \right\} \quad (2.11)$$

where
W/S is the wing loading (lb/ft^2)
δ is the pressure ratio (pressure at altitude/pressure at sea level)
M is the Mach number

The wing loading for cruise analysis purposes is usually a mid-cruise wing loading. And it can be derived for any value of cruise lift coefficient, speed and altitude combination by rearranging Equation (2.11):

$$W/S_{cruise} = C_L \times \{1481.3\delta \cdot M^2\} \tag{2.12}$$

From Equation (2.11), we see that as we burn fuel and the weight decreases, the wing loading and, consequently, the lift coefficient will decrease. Most jet transports are designed to operate at some optimum cruise C_L range which offers the most efficient L/D value. Let us call this optimum L/D value for the first cruise stage as L/D_1. At the initial cruise altitude, the aircraft will be at a higher C_L value (due to the higher wing loading). This will gradually decrease to the optimum cruise design C_L as fuel is burned. As the aircraft continues to burn fuel, the lift coefficient will continue to decrease until such a point that it is necessary for the aircraft to climb to a higher altitude to get back into the optimum C_L range. In other words, delta (δ) will have to be decreased so that C_L may increase to improve the L/D. Let us call this L/D value for the second higher cruise altitude as L/D_2.

$L/D_1 > L/D_2$, but due to the effect of the higher C_L at the higher altitude, the beginning value of L/D_2 will still be higher than the end value of L/D_1 at the first cruise stage. Figure 2.1 illustrates this point for clarity.

FIGURE 2.1 Lift-to-drag ratio vs. C_L for different altitudes.

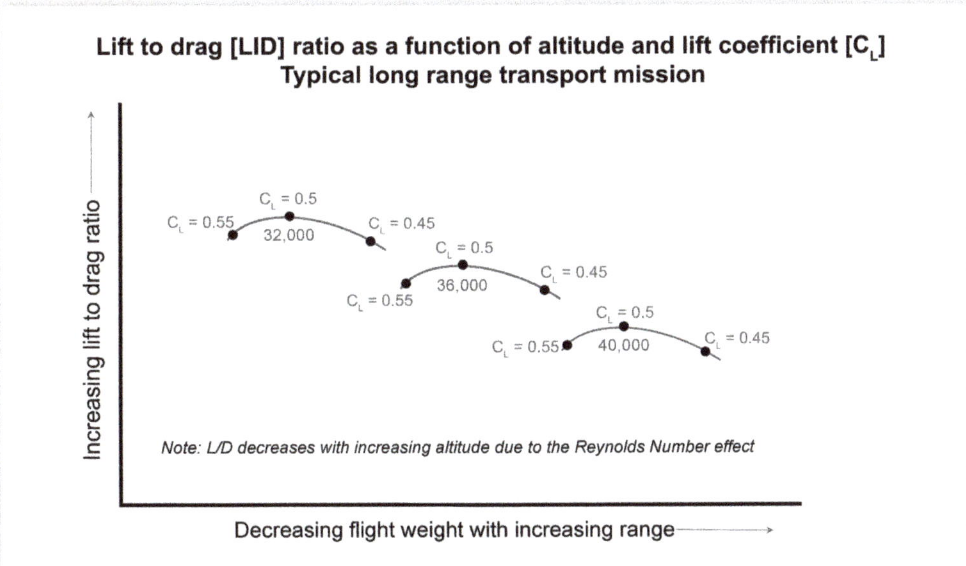

This is exactly what happens in the real world. A jet transport will usually step up in altitude in increments of 2000 to 4000 ft for a long-range mission. This means that if an aircraft starts at an initial cruise altitude of 33,000 ft, as it burns fuel, it needs to climb to maintain its optimum lift-to-drag ratio and fuel burn efficiency. In the course of the cruise mission, it is conceivable that air traffic control permitting the aircraft could step from 33,000 to 37,000 and to 39,000 ft in the final stages of cruise flight.

Given the saturated nature of the air traffic system, however, it is beneficial to set the initial cruise altitude anywhere from 1000 to 2000 ft above the altitude for optimum design C_L. This will ensure that as the aircraft burns fuel while cruising, the optimum altitude will climb (so to speak) to meet the aircraft initial cruising altitude. This technique ensures that there is no immediate need for an early step climb to a higher flight level as this may or may not be permitted by air traffic control in a congested airspace environment.

Induced Drag or Lift-Dependent Drag

Induced drag or drag due to lift is the drag generated as the wing produces lift while moving through the air.

$$CDI = K' C_L \tag{2.13}$$

where C_L is the design lift coefficient.

$$K' = 1 \div (3.142 \times AR \times E) \tag{2.14}$$

where

AR is the aerodynamic aspect ratio = $span^2/S_{ref}$

E is the Oswald's efficiency factor {function of aspect ratio, airfoil technology, and span loading}

Several equations exist that can calculate this factor. However, for transonic wing designs, several factors affect the span efficiency factor, such as wing-mounted engines and handling characteristics. Hence, for this early stage of design, the following equations can be used to determine E:

$$e = 0.97 \times \left\{1 - 2 \times (D_f / b)^2\right\} \tag{2.15}$$

where
D_f is the fuselage diameter
b is the wing aerodynamic wingspan
e is the wing span-loading

$$E = e \times \text{cosine}\left(\Lambda_{c/2}\right)^{0.22} \times \text{SQRT}\left[1 + K_{\text{technology}}\right] \quad (2.16)$$

$K_{technology}$ is the technology factor applied for various aerodynamic technologies

Compressibility Drag

Compressibility drag is also known as wave drag. This is the drag caused by shock waves that form on the airplane. These shock waves are present even at high subsonic speeds, and they can cause separation, which, in turn, causes a drag force.

At this point, it is important to introduce two other very important speeds that have a bearing on the cruise performance of a jet transport. These speeds are the critical Mach number, M_{crit}, and the drag divergence Mach number, M_{div}. The critical Mach number is defined as the Mach number at which the local flow on the wing first approaches a sonic velocity and shock waves begin to form. The drag divergence Mach number (M_{div}) is defined as that Mach number at which the drag coefficient is 20 counts above the critical Mach number, M_{crit}. This is the Boeing definition. The McDonnell Douglas definition of the drag divergence Mach number is that Mach number at which the slope of $\delta C_D/\delta M = 0.10$ [60]. Both definitions give very similar values of the drag divergence Mach number.

$$C_{D\text{wave}} = 0.0013 \times \left[M_{\text{cruise}} / M_{\text{div}}\right]^{13.662} \times \left\{\text{Cosine}\left(\theta_{c/4}\right)\right\}^3 \quad (2.17)$$

Efficient range flying requires the aircraft to cruise at a Mach number that is approximately 0.001 to 0.002 below the drag divergence Mach number. This requires the aircraft to cruise into the drag rise at an L/D below L/D max; however, this cruise condition (L/D cruise < L/D max) offers the best range performance for a jet aircraft.

***Note*:** The most efficient cruise condition for a jet transport is a function of the product of the Mach number and L/D or M.L/D and not just the maximum L/D.

FIGURE 2.2 Graphic showing M_{crit} and M_{div}.

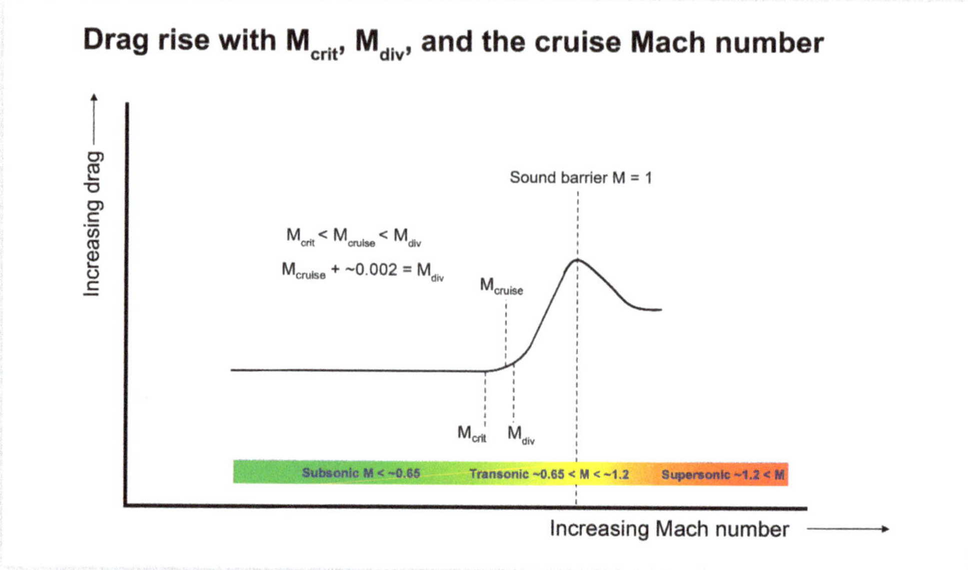

The Weight Ratio

The final term in the Breguet range equation has to do with the weight of the aircraft. This is the ratio of initial weight ($Wt_{initial}$) to final weight (Wt_{final}) of the aircraft in the cruise stage. This ratio while at first glance appears to be unrelated to the various other terms of the Breguet range equation, in reality, is the fall out of the choices that the designer makes for cruise Mach, cruise altitude, cruise TSFC, and cruise L/D. This is not often emphasized, but it is very important that the designer understands the interplay between the various elements of the range equation to obtain a well-balanced high-performance aircraft.

A few key relationships to keep in mind are as follows:

- TSFC α (1/Weight): Typically, engine fuel burn efficiency is directly proportional to the bypass ratio (BPR). A high BPR is synonymous with a lager fan (all else being equal) and a larger nacelle diameter. An increased fan diameter and nacelle diameter lead to a heavier engine, which requires a stronger strut/pylon to hold onto it. A larger nacelle also causes more profile and interference drag, both of which have negative effects on L/D.
- V α Weight: An increase in speed is usually accompanied by a wing, which can be both more highly swept and/or with a lower thickness. Both these wing attributes lead to a higher weight wing structure. Furthermore, a higher cruise speed has a direct impact on the dynamic pressure (at a given cruise altitude), and as a result, the air loads acting on the airframe are higher. The increased aerodynamic loads typically require a larger and hence heavier actuator to move a given control surface. Hence, the structural weight of the aircraft increases proportionally to the speed of the vehicle.

The choice of L/D has an influence on the initial and final weights of the aircraft. The lift-to-drag ratio is related to the aspect ratio, AR, by Equation (1.9), and for a given wing area, the higher the aspect ratio, the higher the span and the lower the span loading. This results in a lower induced drag and, consequently, a higher L/D. However, for a given wing sweep,

a higher aspect ratio wing will typically have a structural wing box, which has shorter chords with less thickness (for a given Mach number).

The reduction in wing area for a given span or more span for a given wing area can both lead to a higher aspect ratio wing. Consequently, strengthening of the wing box structure is necessary to accommodate the subsequent higher torsion and bending loads associated with a more slender wing. This strengthening of the box structure is accomplished by adding excess material, which, in turn, adds weight to the structure of the wing. This is true for a given structural material, whether aluminum or composites. High-aspect ratio swept wings also present unique challenges in integrating wing components such as control surface actuators, torque tube drives, and wiring. These integration challenges add cost and complexity and are difficult to account for in the early sizing phase, resulting in unanticipated weight and drag later in the design process.

Modern airliners from the late 1970s onward began employing what are now called supercritical airfoil sections. This type of airfoil section permits favorable lift-to-drag aerodynamic ratios at transonic speeds. The supercritical airfoil also allows for a thicker wing section (for a given cruise Mach number and sweep), and it carries its peak pressure distribution further aft along its chord. This aft loading increases the structural loading on the flight controls and the power required to hold and move these flight controls. Moving these flight controls under increased load requires bigger actuators and torque tube drives, all of which come at the expense of weight and drag. Aft-loaded airfoils also tend to increase the torsional loading on the wing box, further adding weight to the wing box structure.

For a constant aspect ratio, an increase in the lift-to-drag ratio is usually achieved via a larger wing, which leads to a higher wingspan. However, due to the increased wing area and span, the structural weight of the aircraft will also be higher.

The obvious benefit of the higher aerodynamic efficiency is its beneficial impact on the fuel burn, which, in turn, translates to a lower MTOW (final weight). As a consequence, the aircraft designer is always chasing an optimum balance between aerodynamic efficiency, structural efficiency, and cost.

The Aircraft Sizing Process

So where do we begin? As mentioned in Chapter 1, every aircraft design begins with a top level marketing requirement that describes the necessary attributes for the aircraft to be successful in the market. This is then converted into a set of airplane-level requirements.

Note: A requirement is that attribute that must be met by the aircraft design while an objective is one, which is a nice to have (from an increased marketability, performance and design standpoint), if it can be achieved at a minimum impact to the aircraft's performance and value.

Example of a subset of airplane-level requirements

- Entry into service (EIS)
- Dual-class passenger count
- Design range
- Cargo
- Take-off field length (TOFL)

- Approach speed (Vapp)
- Landing field length (LFL)
- Cruise Mach
- Initial cruise altitude capability (ICAC)
- Distance to initial cruise altitude
- Block fuel per seat at economic range (relative to a base model)
- Cost per seat at economic range (relative to a base model)

During the sizing process, it is beneficial to keep the requirements of all members of the aircraft family in mind when deciding upon the wing area and fuel volume requirements. There is also a weight and performance benefit to this approach in designing an aircraft. This rationale is based upon the fact that historically, when a new aircraft is designed, most of the engineering design community (aerodynamics, loads, structures, systems, etc.) will unwittingly design some margin into the aircraft structure. This margin will be in excess over the already included factors of safety that are required by the certification authorities. The practice of designing margin into the structure and/or aerodynamics is not intentional but a byproduct of the design process that is still dependent on the engineer's knowledge, expertise, and engineering judgment. Hence, most initial models typically end up with some structural margin in various areas. The added margin of structural strength can then be taken advantage of to grow the initial model with minimal structural gage ups and hence a minimum weight penalty.

The advantage of this approach is that after the first model of the type is flying, an enormous database becomes available to the engineering team. This database can then be used to further improve the follow-on family member, giving the later models perhaps a better capability than initially envisioned. This can also be very profitable for an aircraft manufacturer. Why? Because aircraft manufacturers carefully choose the payload-range capability to make sure that they are able to capture a large percentage of the airlines market needs while minimizing the variability in model type. This is important for a manufacturer as added variability and complexity are usually accompanied with an equivalent increase in cost and possibly aircraft price.

Take, for example, the Boeing 747-400. This aircraft has a range of 7,250 NMI with a payload of about 370 passengers or about 83,250 lb (at 225 lb a passenger). To do this mission, the aircraft requires a maximum take-off weight (MTOW) of 875,000 lb. However, only a few airlines truly use the 747-400's maximum payload and range capability. So, an airline may buy the aircraft at a reduced price at a gross weight of, let us say, 810,000 lb. This may allow the airline to fly a shorter mission stage length with a different payload. This is beneficial to the airline as the acquisition and operating costs (airport landing fees, cockpit crew operating costs, acquisition price, etc.) for an airline are usually directly driven by the aircraft gross weight (which is a function of payload-range capability). However, this airline cannot legally fly the aircraft beyond the 810,000 lb take-off weight capability at which the aircraft was purchased. A few years down the road as the airline starts to expand its route structure, it may need to operate the aircraft at its MTOW, and consequently, it will have to buy additional gross weight capability up to the aircraft's MTOW of 875,000 lb. This is typically where a manufacturer can make a healthy profit for what is essentially a simple paper change.

This practice is now changing in the face of aggressive competition. Boeing, for instance, is offering the 787 with unprecedented range capability right "out of the box," so to speak.

In spite of teething problems around the airplanes system reliability, the 787 is achieving its payload-range and economic targets, and this strategy has been very effective as reflected in the airplane's sales numbers. The challenge, of course, was for Boeing to design the associated systems and structure without the excessive margin and hence the excessive weight, which is typical of new designs.

With the 787, an airline can choose the required gross weight suitable for a particular route and then at a later date choose to upgrade the aircrafts weight (via a paper change) free of charge. Here, the thinking is that a more standard aircraft (with fewer options) will retain a higher residual value than a more customized one, thereby making the business case stronger for an airline considering this aircraft. Airbus has followed suit in its bid to stay competitive with its American rival.

The designer must bear in mind; however, that designing excessive range capability to capture the needs of a few key customers can also be detrimental to the aircraft's economics and marketability for a segment of the market that may not require the range capability. For example, if the marketing studies indicate that the bulk of the market need for a transport is 6600 NMI at a given payload, then designing beyond this payload-range requirement will exact a penalty in weight and, consequently, fuel burn. This, in turn, may turn away other potential airline customers who are looking for a higher operating efficiency in an aircraft with a range less than or equal to 6600 NMI. Excessive capability in any performance metric, whether fuel burn or range capability, also drives increased development costs for a manufacturer and a consequent increase in product unit cost. All these factors have to be assessed and the penalties well understood before any decision is made to expand the initial capability for a given jet transport.

The Mission

The typical mission for a commercial transport is defined as follows:

Design Mission

- Phase 0—Start up, warm up, and taxi out
- Phase 1—Take off
- Phase 2—Climb to cruise altitude
- Phase 3—Cruise to destination
- Phase 4—Descent
- Phase 8—Landing, taxi in, and shut down

Alternate or Reserve Mission

- Phase 3a—5% contingency fuel of design mission
- Phase 5—Loiter for 30 minutes
- Phase 6—Execute a missed approach and climb to cruise altitude
- Phase 7—Cruise to alternate airport (200 NMI is typically use for generic mission calculations)

Each of these phases will have an associated fuel burn percentage that can be reasonably guesstimated using data from existing aircraft. The cruise and loiter phases, however, will be calculated after defining the aircraft geometry to calculate the lift-to-drag ratio and various other parameters.

Due to the lack of information at this early stage of the sizing process, we define an equivalent mission that will minimize error by placing more emphasis on those stages that are geometry dependent.

The following values work well in estimating the amount of fuel burned as a percentage of maximum taxi weight:

Phases 0, 1, 2, 4, and 8 can be combined and expressed as follows:

$$\text{Phase-other design} = 0.9425 + k \times RF \tag{2.18}$$

$RF = a \times M/TSFC \times L/D_{cruise}$
$k = 1.86 \times 10^{-6}$ for single aisles
$= 1.13 \times 10^{-6}$ for twin aisles

Here,
a is the speed of sound in knots
M is the cruise Mach number
TSFC is the thrust-specific fuel consumption
L/D_{cruise} is the cruise lift-to-drag ratio

***Note*:** The numbers Phase-other are for the design range. Most transports are evaluated at an economic mission range lower than the design range, which is meant to represent a high frequency of mission's typical missions flown. For long range, twin aisle transports, the economic evaluation range can be 3000 or 4000 NMI. For short range single-aisle transports the economic evaluation range can be 800 or 1000 NMI. For these economic mission evaluations,

$$\text{Phase-other econ} = 0.9594 + k \times RF \tag{2.19}$$

$RF = a \times M/TSFC \times L/D_{cruise}$
$k = 1.12 \times 10^{-6}$ for single aisles
$= 9.37 \times 10^{-7}$ for twin aisles
Phase 3 = to be calculated

Alternate or Reserve Mission

$$\text{Phase 3a} = 1 - 0.05 \times \{1 - (\text{Phase-other design} \times \text{Phase 3})\} \tag{2.20}$$

Phases 5 and 6 can be combined and expressed as follows:

$$\text{Phase-other reserve} = \text{Phase}(5\,\&\,6) = 0.9226 + k \times RF \tag{2.21}$$

$RF = a \times M/TSFC \times L/D_{cruise}$
$k = 3.48 \times 10^{-6}$ for single aisles
$= 3.28 \times 10^{-6}$ for twin aisles
Phase 7 = to be calculated

What remains is to calculate the weight fractions for the cruise stages (design range and cruise to alternate airport) and the loiter stage. We start by defining the aircraft geometry so that we can extract the necessary data required to obtain the lift and drag characteristics of the aircraft.

Define the Fuselage Cross Section

The choice of the cross section is typically governed by the family plan. The family plan is driven by the minimum and maximum seats required to satisfy the market space, which, in turn, can drive the length of the airplane once the seats abreast in economy are known. A lower number of seats abreast results in a longer airplane for a given seat count, and vice versa. Maximum length is an important parameter because we typically want to limit the fuselage fineness ratio (the ratio of the fuselage length to its diameter) to no greater than 13. The Airbus A340-600 has a fineness ratio of about 13, and the B757-300 has a fuselage fineness ratio of 14.3. The A340-600 suffers from excessive body flexing (longitudinally), which requires active damping to keep the bending modes to an acceptable level. This is necessary for technical reasons as well as passenger comfort. Increased fineness ratios can also add significant weight to the fuselage due to strengthening of the fuselage skins, stringers, and frames. Integrating active systems to resolve structural issues adds complexity and cost, with the possibility of affecting reliability. Furthermore, a fuselage that flexes excessively can create issues with regard to ducting, drainage, and other transport elements that run the length of the aircraft.

In the absence of a cross-section drawing, the following equations and approximate dimensions work well and can be used to derive an approximate cross-sectional diameter (Tables 2.2–2.5):

$$\text{Total diameter}\,(\text{in.}) = \left[\text{Inner diameter} + 2 \times \text{Frame depth}\right]$$

TABLE 2.2 Definitions and equations for a fuselage cross section.

	Economy	Business class	First class
No. of seats abreast	SAE	SAB	SAF
No. of aisles on the main deck	AMD	AMD	AMD
No. of aisles on the upper deck	AUD	AUD	AUD
Aisle width (in.)	AWE	AWB	AWF
Seat width (in.)	SWE	SWB	SWF
Inner diameter (in.)	= [SAE + (AMD +1)] × 2 + [SAE × SWE] + [AMD × AWE] + 3	= [SAB + (AMD +1)] × 4.75 + [SAB × SWB] + [AMD × AWB] + 3	= [SAF + (AMD +1)] × 5.0 + [SAF × SWF] + [AMD × AWF] + 3

TABLE 2.3 Seat widths for twin-aisle and single-aisle transports.

SEAT width (in.)	
SWE	17.0 (777-300ER 10-abreast configuration)
	17.2 (B737, B747, B757, B777-9 10-abreast, B787 9-abreast)
	17.7 (A350-XWB early press reports)
	18.0 (A320, B767, A330, A340, A380)
	18.5 (787 8-abreast, 777-300ER 9-abreast)
	19.0 (premium economy)
SWB	19.0 (737, A320, 767, etc.)
	20.0 (A330, A340)
	20.5 (B777, B787, etc.)
SWF	21.0 (B777, B747, B787, A380, etc.)

TABLE 2.4 Aisle widths for single-aisle and twin-aisle transports.

AISLE width (in.)	
AWE	16.5 typically used by charter carriers
	17.0 is marginally acceptable
	18.0 is good design practice
	17–23 is the aisle width range for most narrow body transports
AWB	18.0 or greater (good design practice)
AWF	19.0 (good design practice)
	22.0–27.0 typical fall out values

TABLE 2.5 Approximate fuselage frame depths based on cross section types.

Frame depth
$FD = 1.92 \times e^{(\text{Inner diameter} \times 0.0046)} \times K_1 + K_2$
K_1 = 1 for single-deck airplanes
= For double-deck airplanes If (seats abreast upper deck/seats abreast on lower deck) > 0.4 then
= {(seats abreast upper deck/seats abreast on lower deck)/0.4}$^{1.3}$
K_2 = 0 for single-deck airplanes
= 1 for double deck
K_1 and K_2 are factors used to capture the affect that the upper deck floor load has on the frame depth. Particularly to account for crash loads

Since both a 17.2 in. seat and an 18 in. seat are acceptable as evidenced by decades of jet transport flight, if one takes the average of the two seat widths in the economy section, you get a 17.6 in. seat bottom, which is probably a good start for an acceptable comfort level.

To be clear, the term "comfort level" is rather subjective and should be used sparingly. Several studies have shown that in the economy section, passengers may prefer to give up seat width if you gave them a bit more leg room. These studies also show that it is the spaciousness at the seated eye level that plays into the perceived comfort level of passengers. A good case in point is the Boeing 787 which has continued to sell well and has a 17.2 in. seat bottom in a nine-abreast configuration. The A350-XWB cross section can accommodate a 9-abreast configuration with 17.7 in. seat bottoms.

The cross section for a given configuration can be driven by either the economy class layout or business class. On rare occasions, it could be driven by a first-class layout. Once we have the diameter, we need to select the effective fuselage length. This can be done easily based on the simple metric of floor area per passenger derived from a survey of different airline cabin layouts and manufacturer seating claims. The following values can be used (Table 2.6).

TABLE 2.6 Floor area per seat for twin-aisle transports.

Equivalent seats abreast (ESA)	ESA = SAE × $(SWE/17.2)^{1.6}$ × $(AWE/18)^{0.9}$
Floor area per seat—tri-class seating	= [14.8 − 0.1133 × (PAXMD/ESA)] × K_b × K_s
Floor area per seat—dual-class seating	= [11.01 − 0.0414 × (PAXMD/ESA)] × K_b × K_s
Business class correction factor (K_b)	= $(6/SAB)^{0.15}$
Staircase correction factor (K_s)	= 0.95 for a single stair case (B747-8)
	= 0.84 for a dual stair case (A380)

Table 2.7 estimates the square footage per passenger for typical passenger seat counts in tri-class and dual-class configurations based on average seat counts based on representative airline seating arrangements. Airline seat counts can vary based on their cabin product type and city pairs being flown. Hence, every effort should be made to understand the airline customer's needs so that the fuselage cross section and body length allows for flexibility and maximum market capture.

TABLE 2.7 Floor area per seat for single-aisle transports.

Floor area—dual class seating	= 0.00036 × K_{sb}^2 − 0.0999 × K_{sb} + 13.38
	Here K_{sb} = SAE × $PAXMD^{0.6}$
Floor area—single class seating	= 5.25 + $(6/SAE)^{0.55}$

While this method allows us to set the fuselage length for the respective aircraft, we still need to predict a reasonable seat count on the upper deck for double-deck aircraft such as the Boeing 747-8 and the Airbus A380.

For both these types of aircraft, a reasonable approximation for the number of seats on the upper deck are as follows:

Double-Deck and One-and-Half-Deck Aircraft

$$PAXUD = K_D \times (SABUD \div SAEMD) \times PAXMD \quad (2.22)$$

where
PAXUD is the no. of passengers on the upper deck
PAXMD is the no. of passengers on the main deck
K_D = 0.14 for a one-and-half-deck configuration (e.g., B747-8)
= 0.5 for a double-deck configuration (e.g., A380-800)
SABUD = Seats Abreast in Business on the Upper Deck
SAEMD = Seats Abreast in Economy on the Main Deck

Note: In the case of the 1.5-deck and 2-deck aircraft, the process requires us to first put in a seat count for the main deck, while the number of corresponding seats on the upper deck will be a fall out calculation.

To define the total fuselage length, we make the assumption that the fuselage cabin lies only within the constant section with the nose and tail cone being two simple conical sections. In reality, a percentage of the cabin length will lie within the non-constant sections; however, the aforementioned assumption produces relatively accurate results for fuselage lengths and overall wetted areas.

$$\text{Total fuselage length} = L_f$$

where

$$L_f = \{(\text{Floor area per seat}) \times \text{PAXMD} \div ID_{max}\} - K_{fd} + 2.52 \times D_f + K_{fuse} \quad (2.23)$$

where
ID_{max} is the maximum Inner diameter
K_{fd} is the flight deck location factor
= 0 for single-deck airplanes
= 4 for double-deck airplanes with the flight deck positioned on the main deck or mid-deck
= 12.5 for double-deck airplanes with the flight deck positioned on the upper deck
K_{fuse} is the fuselage factor
= 0 for twin-aisle airplanes
= 12 for single-aisle airplanes
D_f is the external diameter of the fuselage

Once the fuselage length is established, we can calculate the fuselage fineness ratio, which is simply the length divided by the diameter. It is important to note that typically, initial models of an aircraft family start out with fineness ratios between 9 and 10, thus allowing the aircraft to be "shrunk" and/or "stretched" without a significant aerodynamic or structural penalty, respectively.

We now have a preliminary definition of the fuselage that should be in the ballpark of the marketing requirement and the only remaining variables for the geometry will be the wing, empennage, and engines.

Engine Selection

In the absence of an engine deck, we will now have to guess at an approximate engine thrust, bypass ratio (BPR), geometry, and thrust-specific fuel consumption. TSFC can be derived by looking at a host of engines in various thrust classes and extrapolating the TSFC based on the entry into service date [47]. This yearly improvement trend can then be further refined to account for the effect of the bypass ratio [48].

In the real world, the design office would have access to a propulsions technology group that would estimate the future engine performance improvements available to the design team. Hence, the designer would have a reasonable understanding of the engine fuel burn characteristics that would be available at the required entry into service. This is as much an art as it is a science, and it typically requires extensive dialogue between the engine companies and the airframe manufacturer. Items that are discussed include TSFC, bypass ratio, engine weight, reliability, maintainability, and future engine growth in thrust.

At this point in time, it will suffice to know the approximate value for cruise fuel burn at the EIS of the aircraft under study.

$$\text{TSFC} = 0.7275 \times \text{BPR}^{(-0.125)} \times 0.9973^{(\text{EIS}-1995)} \times K_{sfc} \tag{2.24}$$

Here,
EIS is the entry into service date

$$\text{If: SQRT}\{\text{Range} \times \text{Total passenger count}\} < 1118$$

$$K_{sfc} = 1.12 \times 0.000107 \times \text{SQRT}(\text{Range} \times \text{Total passenger count}) \tag{2.25}$$

Otherwise, $K_{sfc} = 1$.

This TSFC value reflects an engine with a bleed air architecture. Bleed air is typically used to provide cabin air and also power the various systems on the aircraft. Bleed air is accessed from the engine compressor and typically reduces the mass flow through the engine, thereby affecting fuel burn. In fact, every aircraft (even all-electric aircraft) will usually benefit from some sort of bleed due to the fact that this improves compressor stability when air is "stolen" from around the tips of the compressor blades.

In an all-electric aircraft, only a small amount of energy is obtained from engine bleed as is necessary to improve compressor stability. The remaining power is drawn from the compressor shaft in the form of electrical energy. The challenge associated with most transport aircraft is that there is a limited amount of horsepower available to power flight control and environmental systems in certain flight conditions. This is especially true during top of descent and in the approach to landing flight regime, when there is a high power demand to run the flaps, landing gear, environmental control system, and, if needed, the anti-ice system. The issue with that is high power demand in a condition when the engines are at a relatively lower thrust setting and hence a low fan and compressor speed. The situation can be further complicated if the power requirements have to be met when one engine is inoperative. This is typically the case in the aircraft design process, where requirements are associated with one or more system failure cases. These failure cases, in turn, can drive the generators on the engine to be oversized, which adds weight and complexity.

Furthermore, in an all-electric aircraft, there is the additional requirement for transport elements (power feeders, etc.) that transport the power to different areas of the aircraft to be sized up in geometry. This may require additional volume within the fuselage and wings, which have little volume to spare. Transport elements required for the electric system (such as power feeders) can add significant weight and may negate any TSFC advantage. All these factors must be taken into account when one chooses the aircraft power system architecture. Additional design issues may include volume for cooling systems for electrical boxes and also thermal shielding that may be required to protect metal and composite areas of the airframe from high-temperature power feeders.

The Boeing 787 is the first commercial aircraft to utilize an all-electric system architecture, thereby getting a fuel burn benefit. Typically, this benefit is advertised to be on the order of 1 to 2% in terms of engine-specific fuel consumption. In the aforementioned TSFC calculation, if we assume that the system's architecture is of a bleedless design, we should reduce the TSFC on average by 1.5% based on industry claims.

Wing Sizing

Wing sizing is one of the most critical aspects of the sizing process. Once the fuselage size is fixed, the size of the wing drives the wetted area of the aircraft. The engine thrust (and geometry), vertical tail area, horizontal tail area, and wing-to-body fairing area are all dependent on the wing area.

The wing of a jet transport is a complex piece of engineering design that is required to do the following:

- Lift a given payload over a certain range and at a required cruise speed.
- Provide the volume to carry the required fuel for the mission.
- More often than not, support the propulsion system which is usually mounted under the wing (B777, A380, etc.).
- Structurally, support the main landing gear and provide the volume for stowage of the same.
- Provide enough high lift capability at low speeds and a high aerodynamic efficiency (lift-to-drag ratio) at cruise speeds.

An explanation of the constraints and their effect on the sizing is prudent. Figure 2.3 shows a representative wing planform.

FIGURE 2.3 Wing planform parameters.

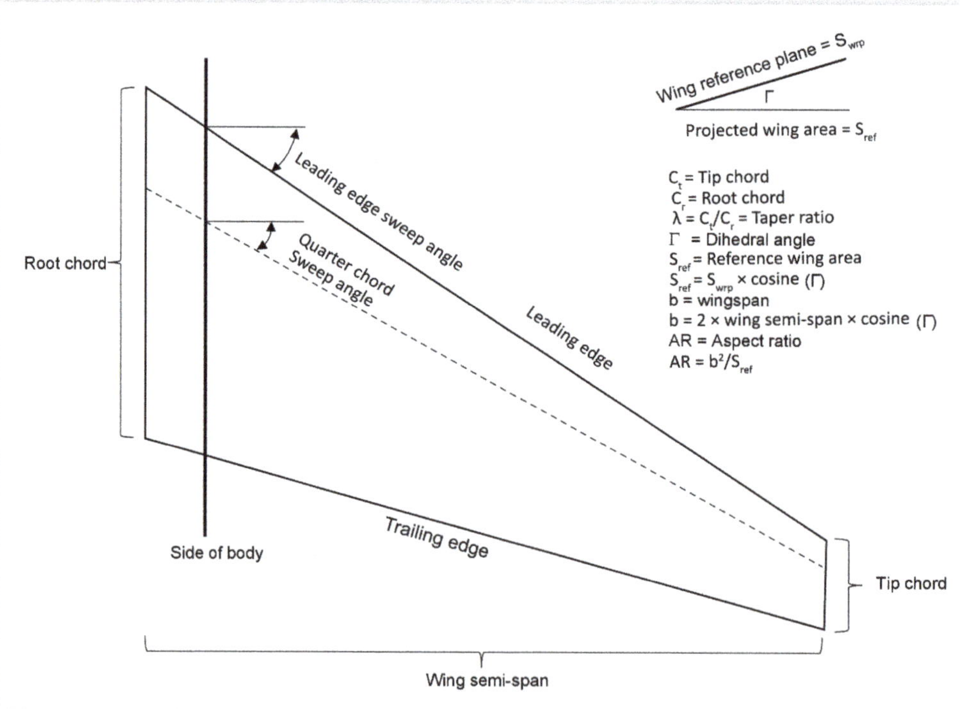

Wing Sweep

Typically, when one refers to the wing sweep, they are talking about the sweep at the quarter chord. This convention will be carried out throughout the book for consistency, unless otherwise specified. Jet transport wings are usually highly swept to accommodate high cruise speeds. This is done (in combination with the thickness-to-chord ratio, t/c) to increase the critical Mach number and delay and/or weaken the strength of the shock wave that forms on the upper surface of the wing.

The lifting characteristics of the wing can be better understood by the lift curve slope. The lift curve slope is defined as the change in the wing lift coefficient per unit change in the angle of attack and is shown in Equation (2.26).

$$CL_{\alpha} = 2.3 \times \{a_0 \times AR / 0.85\} / \\ \{2 + SQRT\left[\left(AR/0.85\right)^2 \times \left(1 - M^2\right) / \left(a_0 / 6.284\right) \times \left(1 + \tan\left(\theta^2 / 1 - M^2\right)\right) + 4\right]\} \times \left(D_f / b_{eff}\right)^{0.25} \tag{2.26}$$

where

$$d_{CL} / d_{\alpha} = CL_{\alpha} = \text{wing lift curve slope in rad}^{-1} \tag{2.27}$$

where
α is the angle of attack between the local free stream and the wing (radians)
θ is the wing sweep at the maximum thickness of airfoil typically around 40% of airfoil length (degrees)
a_0 is the section lift curve slope = $2\pi(1 + 0.8t/c)$
t/c is the average outboard wing thickness-to-chord ratio
Df is the fuselage diameter (ft)
beff is the effective aerodynamic wingspan (ft)
M is the Mach number at which the lift curve slope is being evaluated
AR is the wing effective aerodynamic aspect ratio

It is evident from Equation (2.26) that the greater the sweep of the wing, the lower the lift curve slope. Hence, in a comparison of two aircraft with different sweeps, the one with the higher sweep will need to fly at a higher angle of attack to achieve the same lift. This effect is most pronounced in the low-speed regime typically encountered during take-off and landing. Although the low-speed regime is not usually the one that drives the wing sweep, low speed performance must be assessed to make sure that the aircraft can achieve the desired lift coefficient, especially in the case of aircraft with long fuselages such as the Boeing B777-300ER and the Airbus A340-600, both of which are body attitude limited.

Considerations for choosing the wing sweep are as follows:

High Sweep

- Good high-speed performance (reduces wave drag).
- Not good for low-speed performance.
- High sweep causes an increase in wing weight (compared to a lower sweep wing) due to bending and torsion. Furthermore, a higher wing sweep requires more physical wing structure to achieve the required wingspan necessary to meet the aircraft's performance.
- Sweeping the wing aft places a larger portion of the bending loads on the rear spar. The modern supercritical wing airfoil section typically has its maximum thickness between 30 and 50% of the chord length, and the rear spar is typically located around 65% ± 5% of the chord length. Spar thickness is a key parameter to counter bending loads in a weight efficient manner.
- The combination of high wing sweep with a small taper ratio (Figure 2.3, $\lambda = C_t/C_r$) decreases outboard wing chords. This combined effect forces the outboard wing to operate at higher local cl values, which causes the outboard wing to stall first. This type of wing stall characteristic makes the aircraft pitch up and is a destabilizing effect that must be countered via other measures such as wing twist.

One disadvantage of a high-sweep, high-aspect ratio wing is that the wing can be more challenging to integrate. Integrating the location of the main landing gear on the inboard wing, while trying to provide adequate flap chords, is one of the most challenging aspects of a jet transport wing design. The task is further complicated when one is balancing the design integration of the inboard wing with the high speed demands of the outboard wing (outboard of 40% of the wingspan).

Typically, increasing sweep by ~2.5° increases cruise Mach by 0.01, and vice versa.

For straight tapered wings,

$$LE_{sweep} = \tan^{-1}\left\{\tan\left(\theta_{c/4}\right) + \left[4\times 0.25\times\left(1-\lambda\right)\right]/\left[AR\times\left(1+\lambda\right)\right]\right\} \tag{2.28}$$

$$\theta_{c/2} = \tan^{-1}\left\{\tan\left(\theta_{c/4}\right) - \left[4\times 0.5\times\left(1-\lambda\right)\right]/\left[AR\times\left(1+\lambda\right)\right]\right\} \tag{2.29}$$

Here, AR is the wing aspect ratio.

Thickness-to-Chord Ratio (t/c)

The thickness-to-chord ratio is defined as the ratio of the airfoil thickness to the airfoil chord at a given wingspan station. The choice of thickness-to-chord ratio is driven by several different considerations.

High t/c

- Typically, good for low speed due to a larger leading-edge radius, which improves low speed CL_{max}.
- Increased wing box volume for fuel. Additional fuel in the wing provides wing bending relief and hence a lighter wing.
- Increased wing box depth to efficiently react wing bending and torsional loads, hence reducing wing weight. This is especially important at the side of the body (SOB) and at the inboard engine attachment, both of which see high loads. The downside to high t/c's at the SOB is that for a large SOB wing chord, the increased thickness can create integration challenges between the wing and the fuselage, necessitating a deeper fairing and increased drag.
- For a given wing planform, a higher t/c provides increased volume for systems integration (flap drive systems, spoiler actuators, wire bundles, etc.). This additional volume may be utilized to reduce flap support fairing size by allowing the different mechanisms to be located within the volume between the flap leading edge and the rear spar, as opposed to placing these items in the flap support fairing.
- A high t/c at the engine attach location can provide additional depth for engine attachment. This is especially important in the case of four-engine aircraft, where sufficient depth is required to attach the outboard engine in an area of the wing, where chord lengths are shorter. Typically, at least 16 to 18 in. of box depth is required to allow good outboard engine integration.
- A high t/c at the tip allows for increased depth to attach a wing tip device, if necessary. Box depth at the tip should be at least 6 to 8 in. (the lower number being applicable to smaller transports), depending upon the height of the tip device required.

Low t/c

- Good for high-speed performance due to reduced profile and wave drag. The reduced drag has the effect of increasing the lift-to-drag ratio and hence increasing the range or reducing the required fuel for a fixed range.
- A low t/c at the side of body can result in increased loads, which may necessitate unusually thick wing skin panels, resulting in a weight increase, and manufacturing challenges.
- The efficient packaging of systems may prove to be a challenge when dealing with low t/c's, especially on smaller wing areas (flap mechanisms and other mechanical systems do not always scale down with wing size).
- Typically, decreasing t/c by 1% increases the cruise Mach by 0.01, and vice versa. The high-speed value is primarily concerned with wingspan stations from η = 0.4 to η = 0.9 (or 40 to 90% of the wingspan). An average value can be used for t/c ratios in this region of the wing, or one can simply use the t/c value at 70% of the wingspan as a representative outboard thickness-to-chord ratio.

Aspect Ratio

The aspect ratio is defined as the ratio of the square of the wingspan to that of the reference area. Hence,

$$AR = b^2 / S_{ref} \tag{2.30}$$

where
b is the wingspan
S_{ref} is the wing reference area

Transonic wing aspect ratios usually have values ranging from 7 to 10. In the case of wings with wing tip devices, these values may even get as high as 12, provided the integration issues can be dealt with effectively.

Equation (2.30) defines the wingspan as (b). In most aircraft, the geometric wingspan is also the aerodynamic wingspan. This is also true in the case of tip devices such as raked tips (B777-300ER) which lie in the plane of the basic wing. However, in the case of winglets (A330, ERJ190, 747-400) this is not always true, and the effective aerodynamic wingspan (b_{eff}) is typically larger than the geometric wingspan.

FIGURE 2.4a Airbus A350 XWB Blended Winglet.

Jordan Tan/Shutterstock.com.

FIGURE 2.4b Airbus A380 Wingtip Fence.

Vidit Luthra/Shutterstock.com.

FIGURE 2.4c Boeing 737 MAX Up-Down Advanced Winglet.

Marco Menezes/Shutterstock.com.

FIGURE 2.4d Boeing 777-300ER Raked Wingtip.

Thiago B Trevisan/Shutterstock.com.

FIGURE 2.4e Boeing 777-9 Raked Wingtip in folded position (the first commercial jet transport to offer this feature).

Vidit Luthra/Shutterstock.com.

There are several methods to approximate the effective aerodynamic wingspan (hence the effective aspect ratio) if one knows the dimensions of the winglet. One such method is presented in Figure 2.5.

FIGURE 2.5 Effective aerodynamic wingspan estimation.

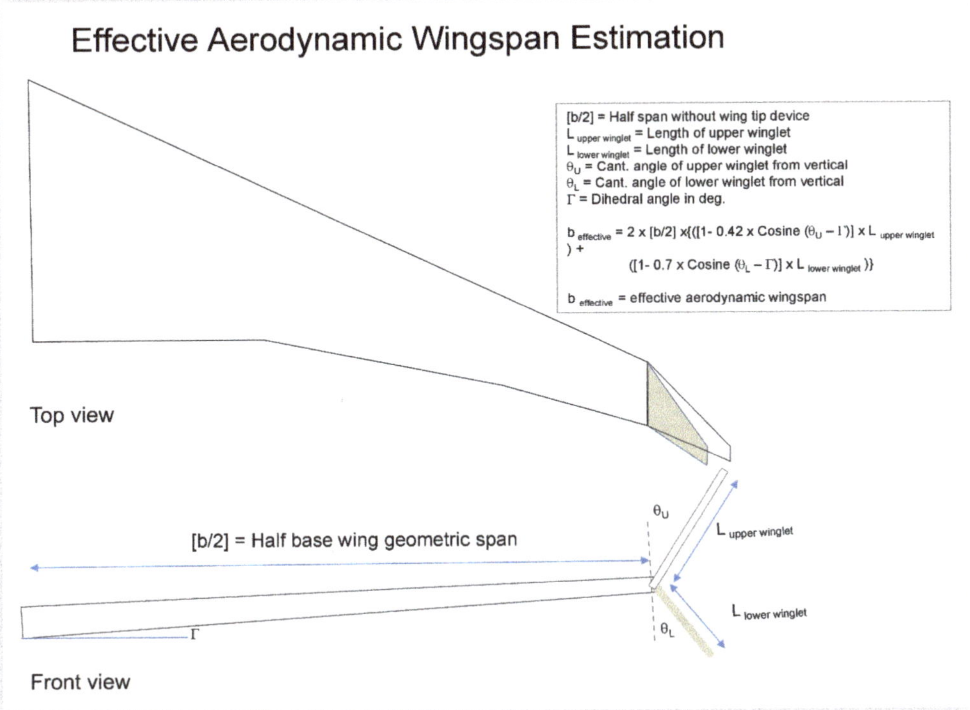

All aerodynamic calculations deal with the effective aerodynamic aspect ratio when not mentioned specifically. The geometric or physical wingspan is only important from the perspective of airport compatibility, where the physical span of the wing can have an impact on parking gates and taxi way maneuvering.

An often-misunderstood notion is that induced drag varies inversely with aspect ratio. Hence, as the aspect ratio increases, the induced drag decreases. However, the induced drag is a function of the wingspan and hence the span loading. So, the higher the span loading (lift divided by the wingspan or L/b = W/b), the higher the induced drag, and vice versa. It must also be emphasized that for a fixed wing area and aircraft weight, increasing the aspect ratio will force the wing to have shorter wing chords and, in particular, cause the outer wing to operate at higher local section Cls, which may increase the wave drag. The increase in wave drag due to high outboard local Cls may require additional wing twist (washout—where the tip chord is at a lower incidence angle than the wing chord at the root) to alleviate the problem. This, in turn, can reduce the induced drag benefit that was expected before the change was introduced.

Shorter wing chords also lead to lower Reynolds numbers and a higher skin friction coefficient. That is not to say that a lower aspect ratio is better than a higher aspect ratio, but it is prudent to understand the limitations of high-aspect ratio wings for jet transports. In every airplane design case, an optimum balance between wing area, wingspan, and weight must be sought out such that the combination has a net benefit to the aircraft's performance.

Taper Ratio

The taper ratio is defined as the ratio of the tip chord to the root chord and is represented by the symbol λ as follows:

$$\lambda = C_t / C_r \quad (2.31)$$

where
C_t is the tip chord of the wing
C_r is the root chord of the wing

Taper ratios on transport jets usually range from 0.2 to 0.4. The lower taper ratios (a more tapered wing) have the effect of reducing the wing weight and increasing fuel volume. This is because a more tapered wing usually biases the wing area inboard and also pushes the center of lift inboard for a given twist. Increasing the area inboard (longer wing chords inboard) results in a deeper inboard wing (for the same t/c), which can hold more fuel by volume. Shifting the center of lift inboard has the effect of reducing the wing bending moment at the side of body. This reduction in wing bending moment makes for a lighter wing structure.

As mentioned previously, lower taper ratios usually cause the outboard wing sections to work harder (operate at higher local Cls) and make the wing more prone to the tip stall characteristic. Tip stall on an aft swept wing can lead to unstable pitch up, which is usually tamed by wing washout (where the tip chord is at a lower incidence angle than the wing chord at the root).

Low taper ratios can also have a negative impact on outboard flap, aileron, and leading-edge device integration. This is primarily due to the shorter outboard wing chords, which have to balance the integration needs of the flap, aileron chord, wing box chord, leading-edge device chord, and volume required for actuation systems for various control surfaces.

Conversely, low taper ratios (for a fixed wing area and aspect ratio) also help with landing gear integration by allowing more of the wing area to be distributed inboard. This additional area inboard has the effect of providing additional volume for landing gear integration but also allows the mean aerodynamic chord to move inboard, thereby providing a favorable position of the main landing gear ground contact point relative to the mean aerodynamic chord.

For initial sizing purposes, the taper ratio of most twin-engine transports (wing-mounted engines) may be approximated to be $\lambda = 0.20$. Transports with four engines on the wing will require higher taper ratios like $\lambda = 0.30$.

The rationale for the higher taper ratio for a quad (four-engine transport) is that usually more outboard wing chord is required to integrate an outboard engine. These taper ratios are referenced for an ESDU-type equivalent wing.

Maximum Lift Coefficient

The maximum lift coefficient, CL_{max}, is the maximum achievable lift coefficient (C_L) before the wing stops flying and separated flow dominates the flow regime. CL_{max} should not be confused with the buffet C_L value, which precedes the stall C_L, especially in the high-speed regime.

The maximum lift coefficient in the take-off and landing configuration is typically a function of the airfoil technology, planform layout, and type of high-lift devices. On a swept wing such as that of a high-speed transport, flap chords on the inboard wing are a strong driver of wing lift in the low-speed configuration. This is because on a swept wing, excessive flap chords on the outboard wing increase the nose down pitching moment and require a stronger download on the tail to trim this moment. This additional down force on the tail degrades the total lift increase.

Selecting the Wing Area

Choose a wing area, quarter chord sweep, average outboard t/c, and aspect ratio based on the airplanes that are currently in service. Do not get hung up on what to choose as the process is an iterative one and will require a few iterations to converge to a good solution. This is done easily on a spreadsheet or within a program. A guess on the wing area, span, and sweep can be made based on data from existing airplanes (Table 2.8).

TABLE 2.8 Estimated wing data for jet transports.

Aircraft	Cruise Mach	Wing area (ft^2)	Wing sweep at quarter chord (degrees)	t/c root at the SOB (approximate values)	Geometric aspect ratio
A320-200	0.79	1290	25.0	0.155	9.6
A330-300	0.82	3892	29.7	0.150	10.05
A380-800	0.85	9016	33.5	0.131	7.55
B767-300ER	0.80	3050	31.5	0.14	8.0
B777-200ER	0.84	4605	31.6	0.14	8.68

Given the sweep, we can also approximate the maximum usable cruise C_L of the aircraft with the required 1.3 g margin to the aerodynamic buffet. It is reasonable to assume that the aircraft will see such C_L values at the top of climb after a maximum gross weight take-off. The 1.3 g margin is there to allow the aircraft to maneuver without causing the airflow to separate, thus causing buffeting. ***This value is not to be confused with the optimum cruise C_L of the aircraft, which can be lower.***

$$CL_{use} = AR \times \left[0.056 - 0.1922 \times t/c^{0.667}\right] + 0.751 \times \left[CL_{cruise} \times \cos \Lambda_{c/4}{}^{1.8}\right] + 0.181 \qquad (2.32)$$

$$\text{If } \left[C_L - CL_{cruise}\right] < 0.07$$

Then,
$CL_{use} = CL_{cruise} + 0.07$
AR is the wing aspect ratio without the wing tip device
t/c is the outboard wing thickness-to-chord ratio
CL_{cruise} is the cruise aircraft lift coefficient derived from Reference [72]

$$CL_{cruise} = \left\{\left[0.2081 + 0.0962 \times AR_{aero\text{-}effective}\right] \times \left[1 + t/c^{0.33}\right] \times \cos \Lambda_{c/4}{}^{0.5}\right\} / SQRT\left[AR_{aero\text{-}effective}\right] \quad (2.33)$$

Here, $AR_{aero\text{-}effective}$ is the effective aerodynamic aspect ratio with the wing tip device.

All that is left is to calculate the average outboard thickness to chord ratio, which will be required to complete the wing geometric requirements to meet the cruise Mach capability.

This can be approximated by a modified version of the Korn equation with a C_L correction.

$$M_{div} = A / \cos\left(\Lambda_{c/4}\right) - \left(0.123 \times CL_{cruise}\right) / \cos\left(\Lambda_{c/4}\right)^3 - \left(t / c_{outbd}\right) / \cos\left(\Lambda_{c/4}\right)^2 \quad (2.34)$$

where
A is the airfoil technology factor for modern post-1988 aircraft
Two- and three-Engine aircraft wing- or tail-mounted engines = 0.926
Four-Engine aircraft with wing-mounted engines = 0.914
Four-Engine aircraft with tail-mounted engines = 0.926

These are good approximations to capture the aerodynamic losses associated with integrating engines on the outboard wing.

CL_{cruise} = design cruise lift coefficient
$\Lambda_{c/4}$ = wing quarter chord sweep angle

We now have enough information to go ahead and calculate the wetted area and lift-to-drag ratio of the aircraft.

The wetted area of an aircraft is defined as the total exposed area of the aircraft. This can be approximated by a modified version of an equation initially defined by Laurence Loftin, a NASA engineer who has contributed immensely to the aerospace community in the area of airplane design and sizing [10].

$$A_{wet} = \left\{\pi \times D_f^2 / S_{ref} \times \left(L_f / D_f - C_0\right) + \left(1.73 + 1.83 \times S_t / S_{ref}\right) + A_n / S_{ref}\right\} \times S_{ref} \quad (2.35)$$

where
A_{wet} is the aircraft wetted area (ft^2)
D_f is the fuselage effective diameter (ft)
$= SQRT\ [W_f \times H_f]$
W_f is the fuselage width (ft)

H_f is the fuselage height (ft)
L_f is the fuselage length (ft)
C_0 = 0.40 single-deck single-aisle aircraft (A320, B737)
= 0.108 single-deck twin-aisle aircraft (B777, A350, B787)
= 0.20 twin-deck aircraft (MD-12X, A380-800)
= 0.51 one-and-half -deck aircraft (B747-8)
S_t is Vertical tail + Horizontal tail area (ft^2)
S_{ref} is the wing area (ft^2)
A_n is the Nacelle area (ft^2)
S_t/S_{ref} = 0.332 for advanced second-generation twin-aisle fly by wire aircraft
= 0.425 for advanced second-generation single-aisle aircraft

For airplanes without fly by wire flight controls, these values can be higher by 0.045 as airplanes with conventional stability tend to have larger empennages with relatively smaller control surfaces (or sized by stability) versus fly by wire airplanes that have reduced empennage areas but relatively larger control surfaces (or sized by control). First-generation fly by wire airplanes can have values about 0.02 higher than shown as technology and aircraft designers were both coming down the learning curve.

A_n/S_{ref} = 0.315 for single-aisle aircraft
= 0.335 for twin-aisle aircraft

Note: The aforementioned method for calculating the wetted area of an aircraft is applicable to tubes and wings. If an unconventional aircraft is sized, the process would typically require a rough drawing of the aircraft configuration, which would then be used to generate wetted area data.

Next, we calculate the minimum profile drag coefficient.

The equivalent skin friction coefficient (C_{Fe}) method is used to approximate the entire aircraft profile drag based on the wetted area from data of known transport aircraft. The advantage of the method is that it allows us to calculate the lift-to-drag ratio rapidly utilizing the aircrafts size (wingspan and wetted area) and cruise Mach number.

$$C_{Fe} = 0.003 \times \left[1.82 - 0.091 \times L_N\left(A_{wet}\right)\right] \times \left(1 + 0.144 \times M^2\right)^{-0.65} \times K_{Reynold\ number} \times K_{eng} \quad (2.36)$$

where
$K_{Reynold\ number}$ = {(Reynolds number at 35,000 ft)/(Reynolds number at cruise altitude)}$^{0.2}$
K_{eng} is the factor to account for number of engines
= 1 for a twin engine (e.g., B777, A350)
= 1.03 for a tri-jet (MD-11, L-1011) and for a four-engine configuration (e.g., 747-8, A380-800)

The Mach number used is the cruise Mach number.

Note: Short-range aircraft may or may not be manufactured to be as "clean" as longer range aircraft due to the smaller pay-off in aerodynamic efficiency over shorter ranges. In some cases, such as in the narrow body market, it is likely that the constraints around an increased manufacturing rate and lower manufacturing costs may outweigh the improvement in fuel burn benefit. However, all else being equal, the impact of fuel in the operating cost model (40 to 50%) necessitates that every effort be made to reduce drag and improve fuel burn.

$$CDpmin = \left[C_{Fe} \times A_{wet} / S_{ref}\right] \times 0.8453 \times \left[1 + 60 \times \left(L_f / D_f\right)^3\right] \times \left[1 + 1.2 \times t/c\right] \tag{2.37}$$

C_{Fe} is the equivalent skin friction coefficient
A_{wet} is the total wetted area
S_{ref} is the wing reference area
L_f/D_f is the fuselage fineness ratio (length/maximum diameter)
t/c is the wing average outboard thickness-to-chord ratio

The maximum cruise lift-to-drag ratio has an effect on an aircraft's endurance capability. In our case, the 30-minute loiter phase will require that the aircraft be flown at maximum L/D in order to minimize the fuel burn.

In the early phases of design, the lift-to-drag ratio can be estimated as follows:

$$L/D_{max\ cruise} = 1/\left\{SQRT\left[4 \times \left(CDpmin + K'' \times C_{lmd}^{2}\right) \times \left(K' + K''\right)\right] - \left(2 \times K'' \times C_{lmd}\right)\right\} \times K_{Mach} \times K_{technology} \tag{2.38}$$

where
C_{lmd} is the lift coefficient at minimum drag
C_{lmd} = k/SQRT [1-m^2]

A good approximation for k for a modern transonic airfoil is as follows:

0.0545 for an advanced airfoil with a conventional trailing edge high lift architecture
0.0575 for an advanced airfoil with a spoiler droop trailing edge high lift architecture
0.0525 for an advanced airfoil with a conventional trailing edge high lift architecture but manual reversion flight controls
K″ is the viscous drag due to lift factor derived from airfoil characteristics. For modern transports, a value of 0.0166 works well.
K′ is the induced drag due to lift factor
K′ = 1 ÷ (3.142 × AR × E)
AR is the aerodynamic aspect ratio = $span^2/S_{ref}$
E is the Oswald's efficiency factor {function of aspect ratio, airfoil technology, and span loading}

Several equations exist that can calculate this factor. However, for transonic wing designs, several factors affect the span efficiency factor, such as wing-mounted engines and handling characteristics. Hence, for this early stage of design, the following equations can be used to determine E.

e is the wingspan loading.

$$e = 0.97 \times \left\{1 - 2 \times \left(D_f / b\right)^2\right\} \tag{2.39}$$

where
D_f is the fuselage diameter
b is the wing aerodynamic wingspan

$$E = e \times \text{cosine}\left(\Lambda_{c/2}\right)^{0.22} \times \text{SQRT}\left[1 + K_{\text{technology}}\right] \quad (2.40)$$

K_{Mach} is the Mach capability given sweep and thickness of the wing
= $[M_{div}/M_{cruise} + 0.002\ ]^{0.6}$ if cruise Mach number is more than 0.002 lower than that calculated by sweep and thickness combination
= $[1 + (M_{cruise} - M_{div})/0.1]^{-0.335}$ if cruise Mach number is greater than the calculated Mach capability
$K_{technology}$ is the technology factor applied for various aerodynamic technologies

Technologies such as laminar flow, adaptive trailing edges, smooth paint, drooped ailerons, and divergent trailing edges can improve aerodynamic performance by up to 2.0% in aggregate (in this case, $K_{technology} = 1.020$). However, care should be taken before assuming aerodynamic improvements. It is necessary to understand the cost of implementation. For example, the application of an adaptive trailing edge to improve cruise performance can involve integration challenges associated with actuation systems along with flight control software implementation, which adds complexity and cost. Research related to real world applications should be conducted before assuming such technologies.

Not investing in these technologies, however, can lead to a loss in competitive market positioning. Hence, one should continue to advance the science, although in a disciplined manner, which requires a multidisciplinary approach (aerodynamics, structures, flight control systems, and manufacturing technologies) in simulation modeling and testing.

The cruise L/D for a constant altitude cruise case (typical for a short cruise to an alternate airport) can be approximated as follows:

$$L/D_{\text{Alt Cruise}} = 0.98 \times L/D_{\text{max cruise}}$$

The Breguet equation can now be used to calculate the fuel fraction for the cruise stage of the flight.

We can calculate the Phase 3 fuel fraction as follows:

$$\text{Phase } 3 = e^{\{\text{Range} * \text{TSFC}/[\text{L/Dcruise} * \text{Vcruise}]\}}$$

Similar to the cruise condition for phase 3, for phase 7, we calculate the fuel fraction for cruising 200 NMI to an alternate airport. Here, we will also assume a TSFC that is similar to the cruise value. Also, the cruise speed will be assumed to be $M_{cruise} - 0.2$, which is close enough for an initial estimate of typical diversion cruise speeds. We will also make the assumption that the alternate cruise stage altitude will be 20,000 ft.

$$\text{Phase } 7 = e^{\{\text{Range} * \text{TSFC}/[\text{L/Dcruise} * \text{Vcruise}]\}}$$

Here, $L/D_{cruise} = 0.98 \times L/D_{max\ cruise}$.
We now have the total mass fuel fraction, M_{ff}.
M_{ff} is the product of all flight phases (Phase 0 through Phase 8).
To Recap:

Design Mission

- Phase 0—Start up, warm up, and taxi out
- Phase 1—Take off
- Phase 2—Climb to cruise altitude
- Phase 3—Cruise to destination
- Phase 4—Descent
- Phase 8—Landing, taxi in, and shut down

where Phase-other design = Phases 0, 1, 2, 4, and 8.

Alternate or Reserve Mission

- Phase 3a—5% contingency fuel of design mission
- Phase 5—Loiter for 30 minutes
- Phase 6—Execute a missed approach and climb to cruise altitude
- Phase 7—Cruise to alternate airport (200 NMI is typically use for generic mission calculations)

where Phase-other reserve = Phases 5 and 6.
M_{res} = product (Phase 3a, Phase-other reserve, Phase 7).
Hence,

$$M_{ff} = \text{Phase-other design} \times \text{Phase 3} \times M_{res}$$

The mass of fuel is calculated as follows:

$$\text{Fuel}(\text{lb}) = (1 - \text{Mff}) * \text{MTW} \tag{2.41}$$

Design Weights

At this point, it is appropriate to introduce the important design weights for a commercial transport. The definitions given by Boeing are the ones we will use (verbatim):

- **Maximum Design Taxi Weight (MTW):** Maximum weight for ground maneuver, as limited by aircraft strength and airworthiness requirements. (It includes weight for taxi and run-up fuel).
- **Maximum Design Take-Off Weight (MTOW):** Maximum weight for take-off, as limited by aircraft strength and airworthiness requirements. (This is the maximum weight at the start of the take-off run and can be calculated as 0.9974 × MTW).

- **Maximum Design Landing Weight (MLW):** Maximum weight for landing, as limited by aircraft strength and airworthiness requirements.
- **Maximum Design Zero Fuel Weight (MZFW):** Maximum weight allowed before usable fuel and other specified usable agents must be loaded in defined sections of the aircraft, as limited by aircraft strength and airworthiness requirements.
- **Operating Empty Weight (OEW):** Weight of structure, power plant, furnishing systems, unusable fuel and other unusable propulsion agents, and other items of equipment that are considered an integral part of a particular airplane configuration. Also included are certain standard items, personnel, equipment, and supplies necessary for full operations, excluding usable fuel and payload. Typically, the Boeing definition includes the "tare" weight or the weight of empty containers (LD3, LD2, etc.). The Airbus buildup of the OEW does not typically include the weight of the empty containers.

We now have sufficient information to go to the next step, that is, to confirm by iteration whether we have guessed at the right MTW and the associated OEW for the given mission accounts for the weight of containers in the lower cargo hold.

The OEW_{calc} based on the MTW is given by a modified empirical equation found in a Society of Weight Engineers (SAWE) paper in Reference [14].

$$OEW_{calc} = \left\{K_{seat} \times K_{fuse} \times A_{wet\text{-}f}{}^{1.22} \times Df^{0.16} \times 0.9985^{(EIS\text{-}1975)}\right\} + MTW \times \left\{5\times10^{-14} \times MTW^2 - 5\times10^{-8} \times MTW + 0.2945 + STM \div 100\right\} \times 0.9985^{(EIS\text{-}1975)} \quad (2.42)$$

Here, K_{seat} can be estimated as follows:
1.00 long-range tri-class seating layouts for twin-aisle airplanes
0.995 long-range dual-class seating layouts for twin-aisle airplanes
0.960 regional dual-class seating layout for single-aisle airplanes
K_{fuse} can be estimated as follows:
1.336 single-deck, single-aisle aircraft where seats x range < 325,000 (e.g., E190-E2)
1.240 single-deck, single-aisle aircraft where seats x range ≥ 325,000 (e.g., A320, B737)
1.164 single-deck, twin-aisle aircraft (e.g., A350, B777)
1.195 double-deck aircraft (e.g., B747-8, A380)
$A_{wet\text{-}f}$ is the fuselage gross wetted area (ft^2)

$$A_{wet\text{-}f} = 3.142 \times D_f{}^2 \times \left[L_f / D_f - K_{f\text{-}deck}\right] \times 0.942 \times K_{close\ out} \quad (2.43)$$

where
D_f is the fuselage effective diameter (ft)
= SQRT[$H_f \times W_f$]
H_f is the fuselage height (ft)
W_f is the fuselage width (ft)
Kf-deck is defined as follows:
0.6 for a single-deck, single-aisle aircraft (e.g., B737, A320)
1.4 single-deck, twin-aisle aircraft (e.g., B777, A350)

0.85 deck-and-half aircraft (e.g., B747-8)
0.95 double-deck aircraft (e.g., A380, MD-12)

$K_{close\ out}$ is the aft fuselage close out factor
1.00 conventional tail (e.g., B787, A320)
1.03 blade tail (e.g., B777, A220, A380)

STM = 2.0 (factor used to capture the weight associated with an oversized wing and engine, and structural margin typically built into the first of model type for growth, e.g., B777-200A, A380-800, and A350-900).

STM = 1.0 factor used for the second model type, typically a stretch and/or an increased gross weight airplane, which follows the initial model after two years. This factor can also reflect that the wing and engine of the growth model are still not at their maximum potential limits. Factor of 1 can also be applied for a re-engine of an existing model.

STM = 0.0 (used for the ultimate growth model, which follows the initial model after more than three years and which utilizes the wing and engines maximum capability).

STM = 3.0 when considering the smallest member in a three-member aircraft family, such that the entire family is sized by the largest member. A good example would have been the A350XWB-800. In the case of the A350XWB, if the wing and engines are sized for the -1000 model, the initial -900 model will accept a small weight and drag penalty. However, the -800 model could have a penalty beyond the weight penalty on the -900 for various reasons such as part commonality, which are driven by cost.

FIGURE 2.6 OEW factor STM to account for family member characteristics.

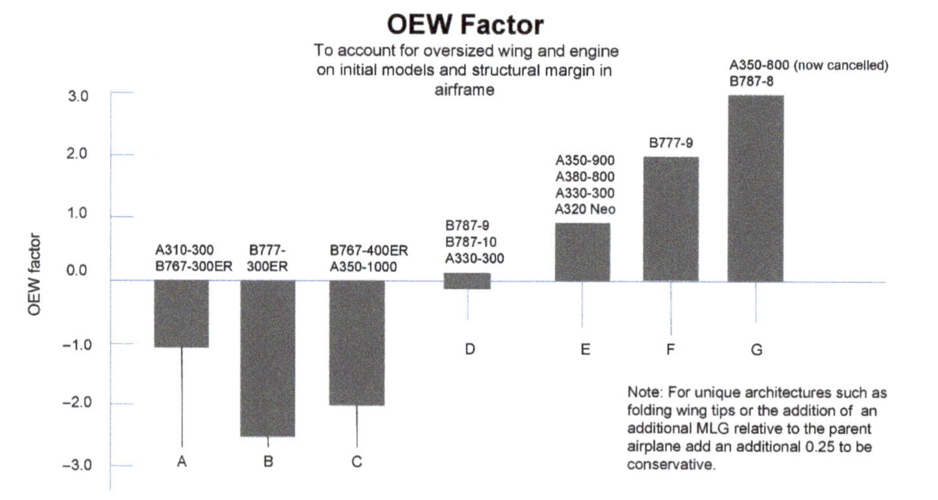

A: Stretch member of the aircraft family with same or modified wing and engine and multiple gross weight (GW) increase
B: Stretch member of the aircraft family with same or modified wing (such as wing tip device), new engine, and multiple gross weight increases
C: Stretch member of the aircraft family with same or modified wing (such as wing tip device), same engine, additional main landing gear and multiple gross weight increases
D: Stretch member of aircraft family with the same wing, and a gross weight increase or initial model at ultimate gross weight capability
E: Initial member of aircraft family (not stretched) with same or modified wing new engine and multiple gross weight increases
F: Initial member of two aircraft family (not stretched) with a new wing, new engine
G: Initial *new* aircraft family member of three-member family with a new wing and engine for future growth. Also applies to a shrink within a two-member family at the same gross weight

However, the fall out OEW is given as follows:

$$\text{OEW} = \text{MTW} - \text{Fuel weight}\left(\text{Wt}_{\text{fuel}}\right) - \text{Payload}\left(\text{Wt}_{\text{payload}}\right) \tag{2.44}$$

where
$\text{Wt}_{\text{payload}}$ is the weight of payload = $N_{pax} \times K$ + cargo weight
K = 225 lb/pax

This is the OEW that falls out as a result of our guessed MTW, mission fuel, and payload requirements.

Once we converge on an OEW, we can now calculate the other critical design weights:

$$\text{MZFW} = K_{\text{aisle}} \times \left[\left(L_f - 2 \times W_f\right) \times W_f \times K_{\text{deck}}\right]^{1.073} = \text{OEW} \tag{2.45}$$

where
L_f is the length of fuselage (ft)
W_f is the width of fuselage (ft)
OEW is the operating empty weight (lb)
K_{aisle} = 20 for a single-aisle aircraft
= 19 for a twin-aisle aircraft
K_{deck} = 1 for a single-deck aircraft (e.g., B777, A320)
= 1.05 for a deck-and-half aircraft (e.g., B747-8)
= 1.14 for a double-deck aircraft (e.g., A380, MD-12)

The MZFW is important because it drives the wing weight and weight of the associated wing support structure. It is calculated using a detailed method that takes into account the maximum number of passengers (single-class passenger count for a single-aisle and dual-class passenger count for a wide-body twin aisle), cargo density, and available volume in the lower hold. The balancing act here is to make sure that the structural capability of the aircraft matches the volume available in the fuselage. Longer bodied airliners such as the B777-300ER and the A340-600 have higher maximum zero fuel weights than their base models due to the excessive cargo volume. This is also the reason why it is speculated that deck-and-half aircraft like the 747 and double-deck aircraft like the A380 have relatively lower maximum zero fuel weights (assuming seat counts remain unchanged). These unique configurations begin to show a benefit as the number of seats increase beyond the point where a single-deck aircraft becomes practical.

The downside of stretching an existing airliner is that the high MZFW also has an impact on the OEW. This is especially detrimental when the stretched aircraft has more cargo volume than is required by the airline market. The presence of additional cargo volume dictates that it may get filled, and hence, the airframe maker has to make sure that the structure has the capability to handle the additional weight. While setting the aircraft price, credit is typically given for cargo carrying capability, and excessive cargo volume beyond the needs of the airline market is simply additional weight, which does not bring a price premium.

At the initial aircraft sizing and conceptual design stage, the MZFW is important as it drives the maximum landing weight (MLW). The MLW is typically set by adding the fuel reserve allowance to the MZFW. At the preliminary design phase, the MLW may be estimated based on historic data as follows:

$$\text{MLW} = \text{A} \times \text{Mission landing weight} + \text{B} \tag{2.46}$$

where
A = 1.06 for single-aisle airplanes
= 1.0495 for twin-aisle airplanes
B = 3200 for single-aisle airplanes
= 36,000 for twin-aisle airplanes

$$\text{Mission landing weight} = \text{OEW} + \text{Design mission payload} + \text{Reserve fuel} \tag{2.47}$$

The Initial and Optimum Cruise Altitude at MTOW

At this point in time, we know enough about the aircraft such as its weights, wing area, wingspan, and fuselage geometry. Our preliminary estimate of L/D is based on the wetted area and the geometry of the wing. However, we do need more insights into the how well the cruise lift coefficient and maximum usable lift coefficient (critical for buffet limited altitude) are matched in terms of the initial cruise altitude capability (ICAC) and the optimum cruise altitude after a maximum gross weight take-off. In real-world aircraft operations, the ICAC and optimum cruise altitude at MTOW are significant in determining how well the aircraft performs compared to its theoretical predictions. To better understand this, it is prudent to introduce the initial and optimum cruise altitude at MTOW definitions and how they affect performance.

The key elements that drive the initial cruise altitude of an aircraft are the take-off weight, aerodynamic characteristics, and thrust capability at altitude of the engines. The maximum usable C_L of the aircraft was determined early in the sizing phase, and now that we have the design weights calculated, we can estimate whether the aircrafts aerodynamics and weight will allow it to achieve the required initial cruise altitude after a maximum weight take-off.

Manipulating Equation (2.11), we can solve for the pressure ratio (δ) as follows:

$$\delta = [\text{W} / \text{S}] \div [1481.3 \times \text{M}^2 \times \text{CL}_{\text{use}}] \tag{2.48}$$

$$\delta = 0.975[\text{W} / \text{S}] \div [1481.3 \times \text{M}^2 \times \text{CL}_{\text{use}}]$$

Here, 0.975 is a factor applied to the taxi wing loading in order to capture the effect of the fuel burn in startup, taxi, take-off, and climb to altitude. CL_{use} is the maximum usable design C_L that offers a 1.3g margin to buffet. This ensures that the aircraft can comfortably maneuver at cruise speeds (using bank angles of 30° or more) while maintaining a safe margin to aerodynamic buffet and the ensuing stall, a method derived from Reference [72].

$$CL_{use} = AR \times \left[0.056 - 0.1922 \times t/c^{0.667}\right] + 0.751 \times \left[CL_{cruise} \times \cos \Lambda_{c/4}{}^{1.8}\right] + 0.181 \tag{2.49}$$

where
AR is the wing aspect ratio without the wing tip device
t/c is the outboard wing thickness-to-chord ratio
CL_{cruise} is the cruise wing lift coefficient

The buffet limited, initial cruise altitude (h_{ICA}) in feet can then be calculated as follows:

$$h_{ICA} = 36{,}089 - 20{,}806.7 \times \ln(\delta / 0.2234) \tag{2.50}$$

Using this method, we can also substitute the optimum cruise C_L value into Equation (2.48) and calculate the corresponding optimum cruise altitude after a maximum gross weight take-off. The optimum cruise altitude should typically be 2000 ft or more below the initial cruise altitude.

As mentioned earlier, engine thrust can also be a limiting factor in ICAC. In fact, for modern jet transports with high-bypass ratio engines, the initial cruise altitude, more often than not, is limited by top of climb engine thrust available.

For operational flexibility and efficiency, it is good design practice to ensure that jet transports have enough thrust to achieve an initial cruise altitude of 32,000 ft, with ultimate growth models achieving 31,000 ft of ICAC. As the aircraft burns fuel (wing loading decreases), the optimum altitude will continue to rise, and when it reaches an altitude 2000 ft above the initial cruise altitude, a jet transport will typically climb another 4000 ft, to once again place itself 2000 ft above the optimum altitude. This step climbing technique ensures that the aircraft is always flying close to its optimum aerodynamic profile. With the latest reduced vertical separation minimum (RVSM) requirements set forth by the certification authorities, the above numbers can be reduced to 2000-ft step climbs and a difference of 1000 ft between the optimum cruise altitude and initial cruise altitude.

FIGURE 2.7 Typical step climb profile.

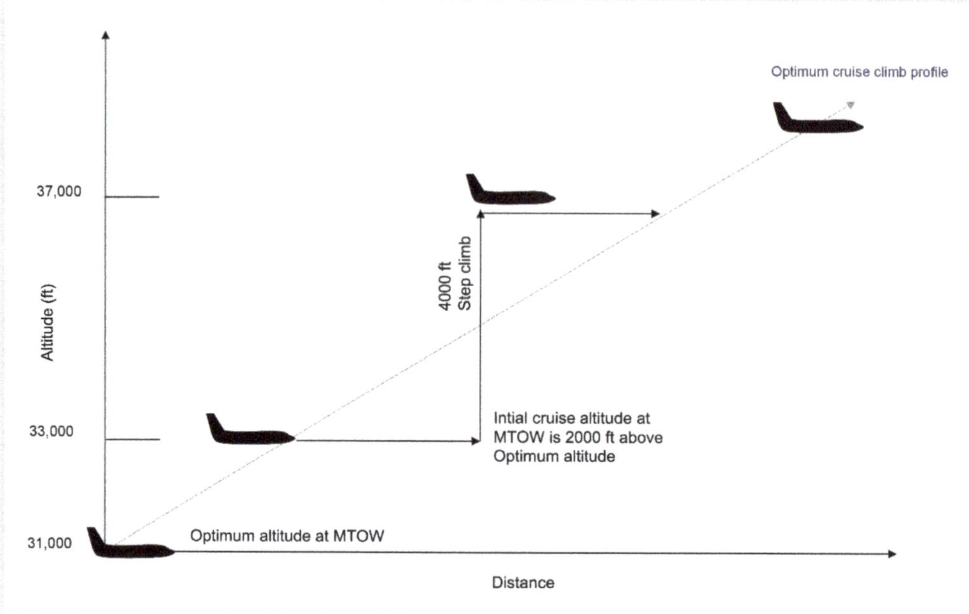

Aerodynamically, the cruise climb profile is thought to be the ideal flight profile where the aircraft always flies at its optimum aerodynamic design cruise lift coefficient by climbing continuously as it burns off fuel. Realistically, engine performance also has to be considered for such a profile to make sure the integrated fuel burn remains optimum.

While step climbing is the operational norm, it is not uncommon for an aircraft to find itself at a nonoptimal altitude for long periods of time due to air traffic control considerations. The introduction of trailing edge variable camber is primarily intended to mitigate the performance penalties in the off-design cruise regions by changing the required aerodynamic shape to optimize the wings performance at the given altitude, weight, and speed combination.

The difference between the optimum cruise altitude at MTOW and the initial cruise altitude should be between 1000 ft (RVSM cases) and 2000 ft. In the case of the A380, the wing is probably sized by the 150 knot approach speed of the A380-900, which is touted as being a 630-passenger stretch airplane. While sizing, if it is found that the difference between the optimum cruise altitude at MTOW and the initial cruise altitude is less than 1000 ft, then the wing area and/or thrust needs to be increased incrementally until the required optimum matching condition is satisfied. For short-range airplanes, it may be acceptable if the optimum and initial cruise altitudes are within 500 ft of each other.

FIGURE 2.8 High-lift device.

High-Lift Devices

Typically, the wing will have some angle of incidence on the fuselage. This is determined by the cruise condition at the top of climb, after a maximum gross weight take-off. The limiting condition here is the fuselage attitude at the top of climb relative to the horizontal, which is restricted to something less than 3°. Two and a half degrees is good design practice. This fuselage angle of attack is selected based on the angle required for minimum fuselage drag and also the maximum slope of the cabin floor that can be comfortably overcome by flight attendants pushing heavy galley carts from the rear of the aircraft to the forward section. This is not necessarily a hard limit on the fuselage attitude, but it is a guideline for good design practice. The DC-10, for example, is an extreme case and has a fuselage attitude (at the top of climb) of ~3.5°.

The swept wing of a jet transport is typically twisted such that the chord at the tip of the wing is at a lower angle of attack than at the root. This is called wash-out, and it is done for various reasons as follows:

- Handling qualities in the stall (forcing the inboard wing to stall before the tips to promote a pitch down attitude after the stall and maintain controllability with the ailerons).
- Tailoring the span load to better balance the induced drag and the wave drag. Tapering the wing planform while good from a structural perspective can increase

the local airfoil lift coefficient (cl) values; these "cl" peaks on the outboard wing can have a negative effect on wing wave drag and also lead to early buffet. Hence, with a more tapered wing, additional washout helps lower the peak cl values and maintain healthy outboard wing aerodynamics.

- Local untwisting is also done in the area of wing-mounted engine nacelles as these can cause a lift loss in the local region of the wing. This is why very often, the wing airfoils decrease in angle from the root to the tip but have a local increase in airfoil incidence in the region of the nacelles to recover some of the lift loss.

At the early stages of preliminary design, the lack of such twist data will force us to make some simplifications and assumptions. We will assume the wing zero lift angle is approximately −0.5° relative to the fuselage longitudinal axis (without flaps deployed). Deploying flaps tends to move the lift curve slope to the left by decreasing the wing zero lift angle (making it more negative).

In most cases, the maximum lift coefficient of a clean swept wing jet transport will lie between 1.3 and 1.6. The lower number is reflective of low-aspect ratio wings and/or wings with higher sweep angles. The higher number is more in line with regional-type jets, which have lower cruise speeds and hence lower wing sweep angles. This is currently where the state of the art is, although there will be exceptions depending upon the aircraft's mission. All that said, if a higher clean wing CL_{max} is achievable by the company's aerodynamics department, then by all means, that capability should be assumed.

Most initial aircraft models of a family will be drawn with take-off aft body contact angles between 12 and 13°. Landing aft body contact angles are typically lower than the take-off angles by about 2° based on typical oleo (shock absorber) extension capability. The aft body contact angle for the take-off configuration is with the main landing gear oleo in the extended position. This is because as the aircraft begins to rotate, the wings begin generating lift such that the weight of the aircraft is transferred from the main landing gear to the wings. Hence, in the take-off attitude, the airplane is already beginning to lift away from the ground, and the angle made by the aft body with the horizontal is taken at the extended gear position. On landing however, the aircraft is moving toward the ground, and when the wheels touch, the weight of the aircraft is being transferred from the wings to the main landing gear, and as such, the aft body contact angle is taken to the compressed main landing gear position.

FIGURE 2.9 Fuselage aft body contact angles.

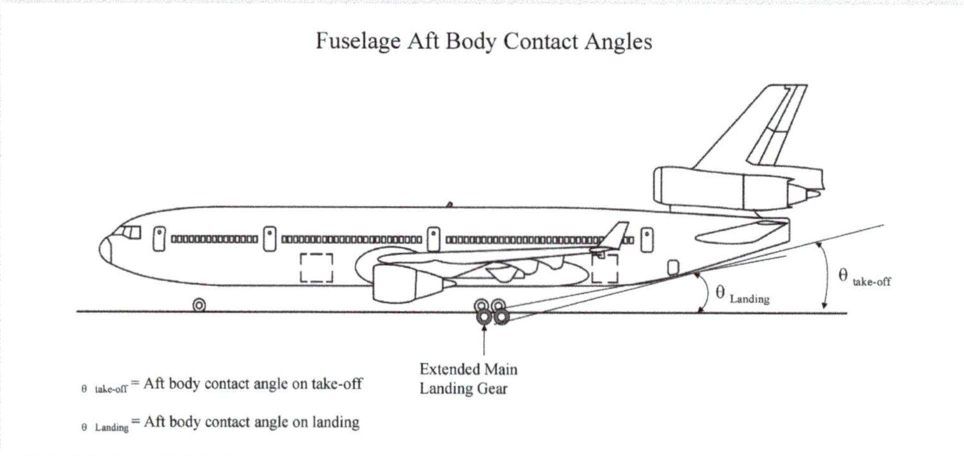

Courtesy of Boeing.

For take-off and landing calculations, however, we will not be assuming the maximum contact angles as shown in Figure 2.9. For obvious reasons, the aircraft's performance must be adequate in the take-off and landing/approach configuration with a margin of safety to ensure that the fuselage does not come into contact with the ground. This contact is often called a "tail strike," and although not very common, it does occur from time to time due to several reasons, some of which are related to airport conditions and pilot technique. The most inherent risk for a tail strike, however, is driven by the geometry of an aircraft, and as such, it is incumbent on the designer to design the airplane such that it is similar in flying qualities (landing and take-off attitudes) to existing transports of the day.

High-Lift Devices and the Lift Curve Slope

The lift of the wing can be increased at low speeds by deploying high-lift devices. High-lift devices typically fall into two categories, namely, leading-edge devices, which are found on the forward or "leading edge" of the wing, and trailing-edge devices, which are found on the aft or "trailing edge" of the wing.

Leading- and trailing-edge devices of a swept wing work together to increase the lift of the wing at low speeds while allowing the aircraft to operate at reasonable angles of attack. In other words, keeping the fuselage attitude at acceptable angles (relative to the horizontal) for passenger comfort, pilot visibility, and safety.

FIGURE 2.10a The lift curve slope.

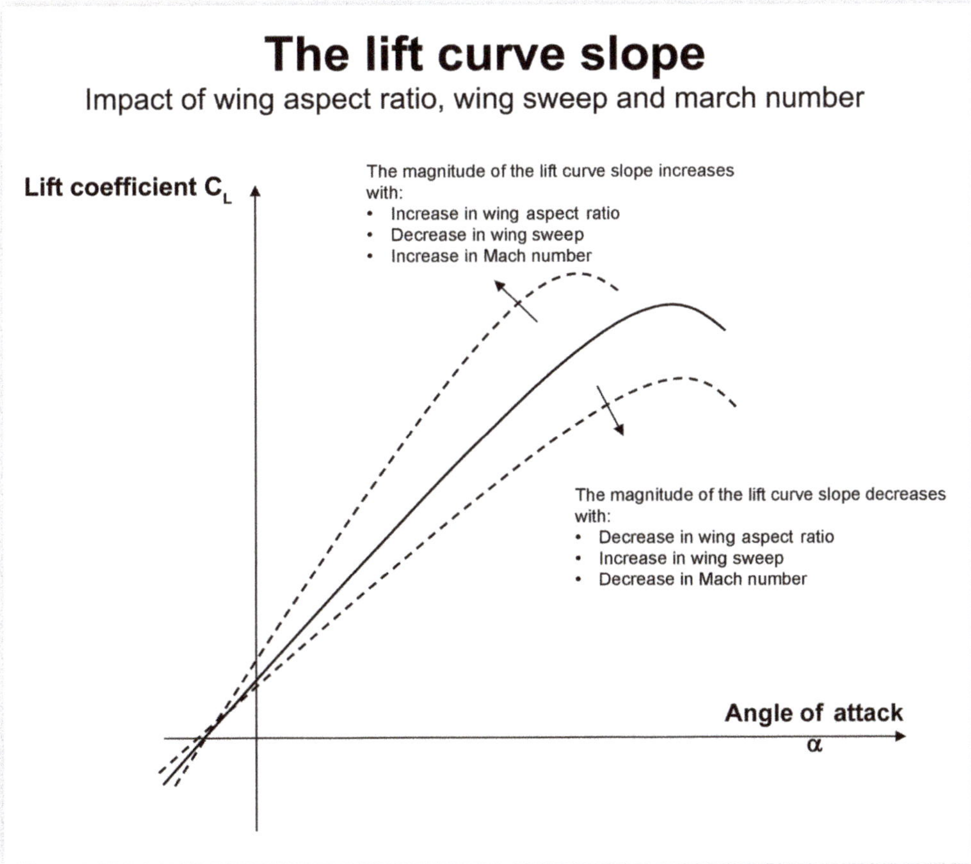

FIGURE 2.10b The lift curve slope and the effect of high-lift devices.

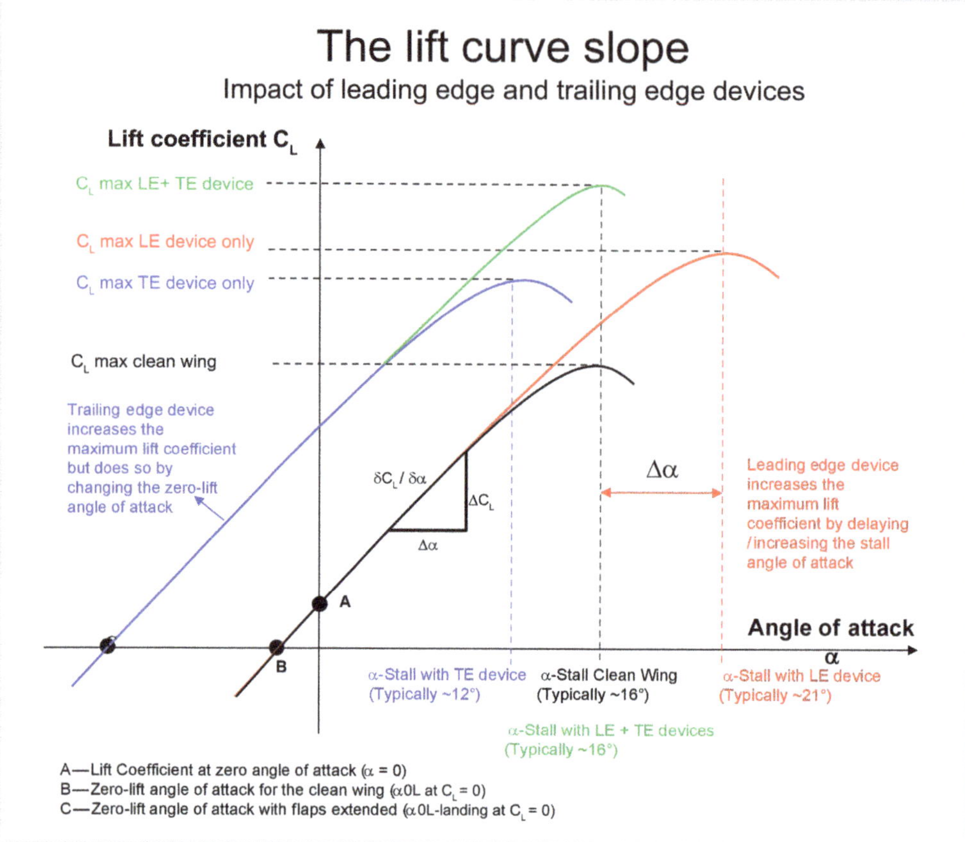

FIGURE 2.11a Leading-edge devices.

FIGURE 2.11b Trailing-edge devices.

As shown in the Figure 2.10b, the leading-edge device allows the wing to operate at a higher angle of attack, while the trailing-edge devices allow the wing to provide more lift at a given angle of attack. Wings with modern airfoils typically stall at an angle of attack between 16 and 18°. Deploying trailing-edge devices or flaps increases lift at a given angle of attack, but they also lower the stall angle of attack, hence the need for leading-edge devices, which allow the wing (with trailing-edge high-lift devices deployed) to increase the stall angle that provides a safe operational margin.

Leading-edge devices can be fixed or movable. Early jet transports such as the de Havilland Comet, Douglas DC-8, and Lockheed Starlifter did not have moveable leading-edge devices, but instead had fixed leading edges that were shaped to provide a compromise between low-speed and high-speed flight.

Early Boeing Jet transports like the B707 had rigid Krueger leading-edge devices. The Boeing 747 pioneered the variable camber Krueger device, which has a radius to the leading edge. Jet transports today have moveable slats on the outboard wing, with a variety of different inboard (inboard of the engine) devices that range from simple droop flaps, Krueger flaps, and/or slats.

***Note*:** Variable camber Krueger (VCK) devices typically perform at a similar level to slats. VCKs offer improved lift due to the ability for low-speed aerodynamicists to tailor the leading edge radius of the device. Typically, the two-position (deployed in the same position for both take-off and landing and retracted for cruise) Krueger devices show a decrease in the take-off L/D value. This is because of the increase in take-off drag as Krueger flaps are not typically sealed on take-off. Three-position Krueger devices (deployed in different positions for take-off and landing, and retracted for cruise) are more challenging to design but offer better performance. Three-position Krueger flaps are still to find their way onto a modern commercial transport. VCKs are not typically found on twin-engine transports, which tend to be second-segment climb drag limited and need to maximize their lift-to-drag ratio in this condition. Quads (or four-engine transports) usually have to meet a lower climb gradient for the second-segment stage and are good candidates for VCK-type leading-edge devices.

Moveable trailing-edge devices also come in different configurations as follows:

- Plain flaps.
- Single-slotted hinged flaps with limited fowler motion (e.g., MD-90).
- Double-slotted hinged flaps with limited fowler motion (e.g., MD-11).
- Single-slotted flaps with fowler motion (e.g., A320, A330, A340, and A380).
- Inboard double-slotted + Outbd single-slotted fowler flaps (e.g., B767 and B777).
- Double-slotted fowler flaps (e.g., B737 MAX family).
- Triple-slotted fowler slaps (e.g., B747-400).

Some key parameters affecting trailing-edge high-lift performance are as follows:

- Flap span as a ratio of wingspan (bf/b).
- Flap chord ratio (cf/c) approximated at 40% of the wingspan (0.4 Eta).
- Wing quarter chord sweep $\theta_{c/4}$.
- Spoiler droop where the spoilers are drooped at low speed, thereby allowing a deeper flap deflection to provide additional wing camber and lift at low speed.

Assuming modern design practices, the delta C_L for high lift can be calculated as follows:

In the Take-Off Configuration

$$\Delta CL_{LE}\ (\text{Leading edge device}) = K_s \times \cos(\theta_{c/4})$$

where $K_s = 0.4$ for early estimates [1, 29].

$$\Delta CL_{TE}\ (\text{Trailing edge device}) = 0.85 \times \delta_f \times (bf/b - 0.1) \times \tau \times CL_\alpha \times \left[1 - 0.08 \times \cos(\theta_{c/4})^2\right] \times \cos(\theta_{c/4})^{0.75} \tag{2.51}$$

where
δ_f is the flap deflection at take-off
= up to 20° for single-slotted flaps
= up to 30° for single-slotted flaps with spoiler droop
= up to 30° for double-slotted and triple-slotted flaps
bf/b is the flap span ratio as a function of wing semispan (without a tip device) – 0.10

Hence on a wing, where the flap span is 0.75, bf/b = 0.75 – 0.10 = 0.65.

τ is the correction for flap chord ratio cf/c

=1 – {acos (1 – 2 × (1 – cf/c)) – sin [acos (1 – 2 × (1 – cf/c))]}/π

CL_α is the low speed lift curve slope

The low speed lift curve slope of the aircraft without flaps deployed can be approximated as follows:

$$CL_\alpha = 1.05 * \{a_0 \times AR / 0.85\} / \{2 + SQRT[(AR / 0.85)^2 \times (1 - 0.2^2 + \tan(\theta^2) + 4]\} \quad (2.52)$$

where

$dCL/d_\alpha = CL_\alpha$ is the wing lift curve slope (rad^{-1})

α is the angle of attack between the local free stream and the wing (rad)

θ is the wing sweep in degrees, at the max thickness of airfoil at 30% of airfoil chord

a_0 is the section lift curve slope or 2π

1.05 is the factor to account for t/c effect on the 2D lift curve slope of the airfoil

t/c is the average outboard wing thickness to chord ratio

W_f is the fuselage diameter (ft)

b_{eff} is the effective aerodynamic wingspan (ft)

0.2 is the Mach number at which the lift curve slope is being evaluated

AR is the wing effective aerodynamic aspect ratio

In the Landing Configuration

$$\Delta CL_{LE}\left(\text{Leading edge device}\right) = K_s \times \cos\left(\theta_{c/4}\right) \quad (2.53)$$

where K_s = 0.45 for early estimates [1, 29].

$$\Delta CL_{TE}\left(\text{Trailing edge device}\right) = 0.91 \times \delta_f \times \left(bf / b - 0.1\right) \times \tau \times CL_\alpha \times \left[1 - 0.08 \times \cos\left(\theta_{c/4}\right)^2\right] \times \cos\left(\theta_{c/4}\right)^{0.75} \quad (2.54)$$

where

δ_f is the flap deflection at take-off

= up to 36° for single-slotted flaps (36° is good design practice for initial sizing; however, an aggressive good design can achieve a deflection to 40°)

= up to 45° for single-slotted flaps with spoiler droop

= up to 46° for double-slotted and triple-slotted flaps

bf/b is the flap span ratio as a function of wing semispan (without a tip device) – 0.10

Hence on a wing, where the flap span is 0.75, bf/b = 0.75 – 0.10 = 0.65.

τ is the correction for flap chord ratio cf/c

=1 – {acos (1 – 2 × (1 – cf/c)) – sin [acos (1 – 2 × (1 – cf/c))]}/π

CL_α is the low speed lift curve slope

The low speed lift curve slope of the aircraft without flaps deployed is given from Equation (2.52) as

$$CL_\alpha = 1.05 \times \{a_0 \times AR / 0.85\} / \{2 + SQRT[(AR / 0.85)^2 \times (1 - 0.2^2 + \tan(\theta^2) + 4]\}$$

where
$dCL/d\alpha = CL_\alpha$ is the wing lift curve slope (rad^{-1})
α is the angle of attack between the local free stream and the wing (rad)
θ is the wing sweep in degrees, at the max thickness of airfoil at 30% of airfoil chord
a_0 is the section lift curve slope or 2π
1.05 is the factor to account for t/c effect on the 2D lift curve slope of the airfoil
t/c is the average outboard wing thickness to chord ratio
W_f is the fuselage diameter (ft)
b_{eff} is the effective aerodynamic wingspan (ft)
0.2 is the Mach number at which the lift curve slope is being evaluated
AR is the wing effective aerodynamic aspect ratio

The ΔCL values derived are based upon the premise that the high-lift systems will be designed using the latest methodologies in CFD and wind tunnel testing. In other words, these values are indicative of the state of the art and should be used to predict and set as a target for the low-speed architecture choice.

As mentioned earlier, we need to assume a take-off and landing attitude to determine the aircraft lift coefficient for these respective phases of flight. Without a three-view/general arrangement drawing at this point, we have to make some assumptions. For initial models of aircraft that have fuselage fineness ratios below 11, we can assume that we can safely rotate the aircraft on take-off to a pitch attitude of 9°. For an initial model of an aircraft that is designed to have an aft body take-off rotation contact angle between 12 and 13°, this should provide a safety margin of approximately 2 to 3° to the aft body. By a similar method, we will assume an approach attitude that has the fuselage angled at 2.0° nose up attitude to the horizontal. The corresponding values for aircraft with fuselage fineness ratios of above 11 can be assumed to be 8° for take-off. The same 2.0° attitude will be assumed for approach to landing.

Observing current jet transports such as the A330, A320, B777, and B787, one can make a choice for a modern high-lift system. There are two approaches to choose from over here. The Boeing approach (Douglas and Lockheed also adopted such configurations) typically requires an all speed inboard flaperon, which is used for high lift at low speeds and roll control. There are several reasons for this, which date back to the beginning of the jet age.

Early jet transports such as the B707, DC8, and L1011 all had wings with high sweep back and thin airfoils. This made for flexible wings which were susceptible to aileron reversal at very high speeds. Aileron reversal is the condition where deflecting the control surface has just the opposite effect. For instance, at very high speeds such as those encountered in cruise or a dive, turning the control wheel (or joystick) left would command the left aileron up and the right aileron down. If the wing is flexible in torsion, this could have the reverse affect by twisting the left wing leading edge up and the right wing leading edge down. This would cause the aircraft to turn to the right or reverse its intended direction of roll. Also, the flaps on these older transports were typically double or triple slotted and required large deflections to achieve the desired lift coefficient. This created a situation in the area behind the engines where it was possible that the thrust could impinge on the flap, thereby causing excessive load on the wing structure. To resolve this issue, manufacturers chose to put a trailing-edge "thrust gate" or break in the flaps between the inboard and outboard sections. It was also logical to add a high-speed aileron in this area to avoid aileron reversal problems. It is only recently that Boeing decided to "droop" the high-speed aileron for improved lift at low speed. This was first done by Boeing on the B767. It is common practice these days to deflect the flaperons and ailerons down on take-off and landing.

Airbus, on the other hand, has split outboard ailerons on the outboard wing. The inner one is used for roll control at high speed, and both are drooped for high lift in the low-speed regime. The Airbus philosophy allows a more continuous trailing edge inboard with a fuller inboard span load. However, the flap span is shorter because of the need for two roll devices on the outboard wing. This typically could hurt low-speed take-off performance. It could also be argued that having a high-rate, high-speed roll control device at the wing tips probably comes at a possible weight penalty in order to keep high-speed aileron reversal in check. This design is also helped by newer supercritical airfoil sections, which came into vogue in the mid-1970s. These airfoils offer improved performance with an increased thickness that helps stiffen the wing box structure. Airbus flight control philosophy locks out the ailerons at dive speeds and relies on spoiler authority only for roll control. The Boeing approach, on the other hand, ensures full high-speed flaperon-driven roll authority in all flight conditions up to dive speed. As flight control algorithms advance, these philosophies and approaches are constantly challenged in an effort to reduce weight and improve fuel burn.

The first new-generation transports from Airbus beginning with the A310, A320, etc. did not have close coupled engines, where the engines are mounted close to the wing lower surface. Airbus turned a potential weakness in the high-speed arena into a benefit in the low-speed regime. The lower mounted engines and improved flap aerodynamics helped designers get around the thrust impingement problem (interaction between flap and thrust plume) by adopting simpler single-slotted flaps without thrust gates.

Both philosophies produce acceptable solutions from a low-speed perspective.

The low-speed lift coefficient equation can be approximated as follows:

For take-off,

$$CL_{\text{Take-Off}} = CL_{\text{clean}} + \Delta CL_{\text{LE}} + \Delta CL_{\text{TE}} \qquad (2.55)$$

$$CL_{\text{clean}} = CL_{\alpha} \times K_{\text{flap-extend}} \times \left(\theta_{\text{T-Off}} + 0.5 + K_{\text{AG}}\right) \qquad (2.56)$$

where

CL_{α} is the low-speed lift curve slope

$K_{\text{flap-extend}}$ is the adjustment of lift curve slope for extended flaps

$= \{1 + [cf/c \times K_{\text{fowler}} \times bf/b]\}$

cf/c is the flap chord ratio

K_{fowler} is the percent of flap fowler motion (e.g., 20% fowler motion would have a value of 0.20)

bf/b is the flap span ratio (assume wingspan without a wing tip device if one is present)

0.5 is the zero lift angle of the wing (degrees)

K_{AG} = 0 conventional main landing gear trucks or boggies

= 1.0 main landing gear trucks or boggies which have a device that allows the aircraft to rotate around the aft wheels, rather than the midpoint of the truck, thereby getting a degree more rotation capability. It is typically found in twin-engine stretch airplanes where rotation clearances are limited and initial climb performance with one engine out is critical (B777-300ER, A330).

By similar comparison, the corresponding landing (approach) lift coefficient can be calculated as follows:

$$CL_{LNDG} = CL_{clean} + \Delta CL_{LE} + \Delta CL_{TE} \quad (2.57)$$

$$CL_{clean} = CL_{\alpha} \times K_{flap\text{-}extend} \times \left(\theta_{Approach} + 0.5 + 3.0\right) \quad (2.58)$$

The number 3.0 reflects the 3°-glide slope that is typical of most airport approaches. However, there are airports such as London City that have steeper approach flight paths on the order of 5.5°. This value or any other specified value should be used if it is determined that this would be the sizing condition for a particular type of transport mission. Typically, in the approach configuration, $K_{AG} = 0$.

Note: The C_L values derived above for the take-off and the approach configuration are calculated at the respective normal take-of and approach attitudes and are **not** the CL_{max} values typically quoted in most text books. In other words, these are the maximum achievable C_L values accounting for the margins required either for fuselage attitude contact to the ground or the margins required before the wing stalls—whichever the limitation may be for a given design.

The take-off and landing configuration speeds can now be determined from the following equation:

$$V = SQRT\left\{2 \times K \times (W/S) / \left[\rho \times C_L\right]\right\} \times 0.5925 \quad (2.59)$$

where
W/S is the wing loading in the given gross weight condition. Use MTOW and MLW as required.
C_L is the $CL_{Take\text{-}Off}$ and CL_{LNDG} for take-off and landing, respectively
ρ is the density of air at the sea level standard condition in slugs (ft^3)
0.5925 is the factor that converts feet per second into knots
K = 1.0 for the approach configuration
= 1.07 for the take-off condition

A check should be made at this point to make sure that the lift off speed is below 204 knots or 235 mph, which is typically the maximum rated speed for most modern air transport tires.

Landing distance can be estimated early on as follows:

$$LFL = \left\{118 \times W/S_{LNDG} \div \left[CL_{LNDG} \div 0.661\right]\right\} + 400 \quad (2.60)$$

Note: Brake performance (brake energy or brake force) can be a major contributor to take-off and landing field length and should be considered as the design matures. At this stage, however, the methodology provided above is a good stand in getting the design in the ballpark.

Thrust Sizing

Several factors can affect the thrust sizing of a commercial transport aircraft. These can include the following:

- Take-off performance
    - Take-off field length
    - Initial climb capability up to 400 ft
- Climb performance
    - Top of climb performance at the initial cruise altitude
    - Distance to climb to the initial cruise altitude
    - Time to climb to the initial cruise altitude
- Cruise performance
- Aborted landing performance when one engine is inoperative

In most cases, it is the take-off and initial climb performance that drive the thrust size. The Airplane Design Series by Jan Roskam goes into more detail on the take-off and initial climb requirements that must be met by aircraft being certified under FAR 23 or FAR 25 requirements.

For preliminary estimates, we will limit our assessment to the take-off thrust requirements and the enroute climb thrust requirements. Aircraft that have their thrust sized by top of climb generally do so because:

- They carry a larger wing than necessary (typical of initial models).
- They have more than two engines and have a lower climb gradient requirement and hence a lower thrust requirement in the initial climb segments up to 400 ft.
- They have high-bypass ratio engines that have a higher thrust lapse rate (thrust loss with altitude) than engines with lower bypass ratios.

The distance taken to climb to the initial cruise altitude can be a significant sizing factor. This is because in several airspaces such as those around London and Hong Kong, if the aircraft is not at its initial cruise altitude before it exits the departure airport airspace, it can be at risk of being stuck at a nonoptimal cruise altitude. Using London as an example, once the aircraft is beyond 200 NMI from London, it is handed over to the mainland Europe air traffic control (ATC). Since European skies are congested, ATC rarely reassigns an aircraft that is handed over to it to a higher altitude. In a manner of speaking, the aircraft is then "stuck" at the altitude that it enters mainland European airspace and may be required to fly through most of Europe's busy airspace at a nonoptimal altitude, thereby hurting fuel burn and impacting an airlines bottom line. Typically, less than 190 NMI is a good airplane design requirement for the largest family member (in a multi model airplane family) where the initial cruise altitude can typically be about 33,000 ft. As aircraft and airspace equipment continues to evolve, this situation could change, and this constraint could be relaxed.

Hence, in keeping with good systems engineering practices, requirements should be pressure tested with every new design.

Similarly, the time to climb can also be set as a constraint to meet certain airspace requirements. Most medium- and long-range transport jets are designed to climb to their initial cruise altitudes at MTOW within 30 minutes. For regional jets, this requirement can be 25 minutes due to the shorter stage lengths. This is sometimes driven by airlines that may want the ability to offer a beverage service on short missions of 300 to 500 NMI. Conversely, it is also entirely probable that most regional jets are designed to have good short field length take-off capability and therefore carry excess thrust, which allows them to climb to altitude quicker.

To keep initial sizing simple, we will use the top of climb thrust requirement as a surrogate for the distance to climb and time to climb requirements.

Take-Off Thrust Sizing

The take-off thrust per engine can be calculated as follows:

$$T/W_{\text{Take-Off}} = \left\{\left\{K_{\text{TOFF}} \times W/S_{\text{MTOW}}\right\} \div \left\{CL_{\text{Take-Off}} \times \left[\text{TOFL} + K_{\text{flare}}\right]\right\}\right\} \div NE \tag{2.61}$$

where

$T/W_{\text{Take-Off}}$ is the thrust-to-weight ratio at MTOW

K_{TOFF} is the factor used to capture the different climb gradient requirements for the number of engines

= 30.8 two-engine aircraft (e.g., B777, A330)

= 28.9 three-engine aircraft (e.g., MD-11)

= 27 four- or more-engine aircraft (B747, A380)

W/S_{MTOW} is the wing loading at MTOW (psf)

$CL_{\text{Take-Off}}$ is the take-off lift coefficient [Equation (2.25)]

TOFL is the take-off field length (ft)

K_{flare} is the adjustment for rotation or flare time. Typically, wide-body airplanes can take up to 4.5 sec to achieve the take-off attitude. Single-aisle airplanes will have lower flare times by about 1 sec.

= 0 for twin-aisle aircraft

= 250 for single-aisle aircraft (assuming an average lift of speed of 250 ft/sec)

NE is the number of engines

The aforementioned equation is used to calculate the one engine inoperative take-off performance up to the obstacle clearance height of 35 ft.

FIGURE 2.11c Take-off diagram.

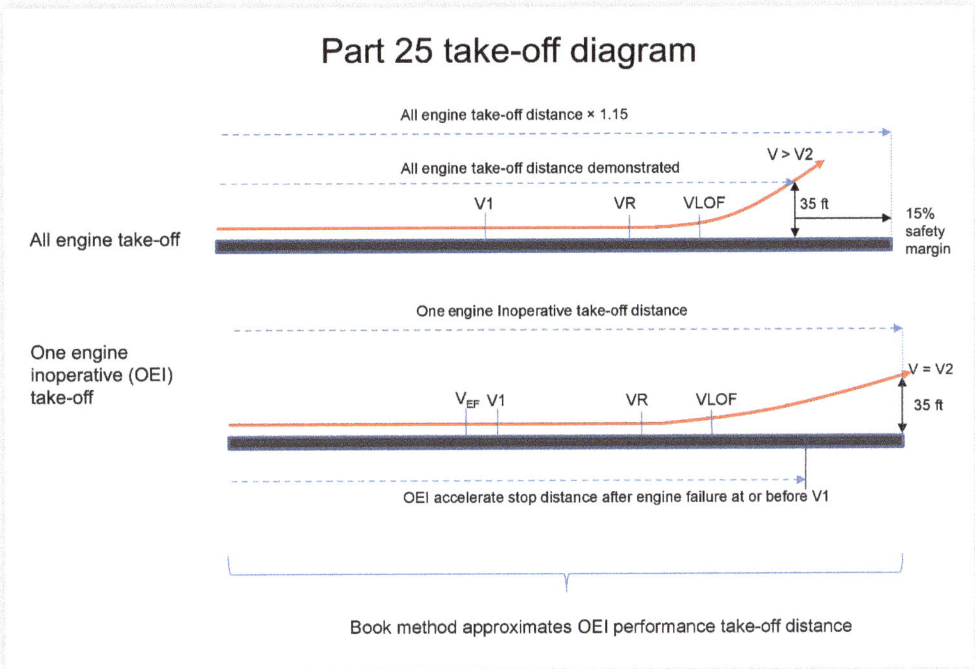

Second Segment Climb Thrust Sizing

Second segment climb is that part of the initial climb out from 35 ft up to 400 ft, where the gear is retracted and the flaps are in the take-off configuration. The calculation is assumed that one engine is inoperative. The climb thrust for this segment of flight can be calculated as follows:

$$T/W_{2nd\ Seg} = NE \div (NE - 1) \times \left\{L/D_{2nd\ seg}{}^{-1} + CGR\right\} \div 0.795 \tag{2.62}$$

where
NE is the number of engines
CGR is the climb gradient required
= 0.024 two-engine aircraft
= 0.027 three-engine aircraft
= 0.030 four (or more), engine aircraft

$$L/D_{2nd\ seg} = 0.96 \times \left\{-159.8 \times K_{CLv2}{}^{3} + 235.6 \times K_{CLv2}{}^{2} - 119.5 \times K_{CLv2} + 24.2\right\} \times (cf/c)^{-0.25} \times (bf/b)^{0.38} \times K_{fd} \times K_{ad} \tag{2.63}$$

where
$K_{CLv2} = CL_{v2}/SQRT\ [AR_{geometric}]$
$CL_{v2} = CL_{Lift\ Off} \times 0.985$
$CL_{Take\text{-}Off}$ is the lift coefficient at take-off
$AR_{geometric}$ is the geometric aspect ratio
cf/c is the flap chord ratio
bf/b is the flap span ratio as a function of wing semispan (without a tip device) – 0.10

Hence on a wing, where the flap span is 0.75, bf/b = 0.75 – 0.10 = 0.65.

K_{fd} = flap drag factor different flap architectures
= 1 single-slotted inboard + single-slotted outboard
= 0.980 single-slotted flaps with spoiler droop
= 0.965 double-slotted inboard + single-slotted outboard (use the same for single-slotted inboard + double-slotted outboard which is a configuration found on the A321neo XLR)
= 0.930 double-slotted inboard + double-slotted outboard

K_{ad} = drag factor on take-off due to aileron droop
= 1.0 for use of aileron droop on take-off
= 0.985 for no aileron droop on take-off

The 0.795 factor is used to represent the ISA + 15°C atmospheric conditions.

Top of Climb Thrust Sizing

Enroute climb thrust requirements can often set the thrust of the engine. The distance required to climb to a specified initial cruise altitude can drive the required sea level static thrust of a jet. It is good practice to not exceed a distance of 200 NMI from the originating airport to get to the required initial cruising altitude.

The time to climb can also set the sea level static thrust required and, hence, the engine size. Most medium- and long-range airplanes have a time to climb threshold of 30 minutes. Short-range jet airplanes which have an average stage length of 500 to 800 NMI typically require the time to climb to be below the 25-minute threshold. The time to climb values mentioned before are the excepted values in the airline industry and are generally good guidelines for climb performance.

The climb thrust requirements mentioned earlier can be captured rather elegantly by understanding the thrust required at the top of climb. The sea level static thrust required to meet our cruise speed at the top of climb is given by the following equation derived from Reference [9].

$$T_{TOC} = \left[0.975 \times MTOW / (L/D)\right] + \left[300 \times 0.975 \times MTOW / (V_{TAS} \times 101.3)\right] \quad (2.64)$$

where
T_{TOC} is the top of climb thrust (lb)
0.975 is the percent of take-off weight after burning fuel in take-off, and climb
MTOW is the maximum take-off weight (lb)
L/D is the cruise lift-to-drag ratio
V_{TAS} is the cruise speed at the initial cruise altitude (knots)

Hence, the equivalent thrust required at sea level static conditions is given as follows:

$$T/W_{TOC} = T_{TOC} / K_{lapse} / MTOW / \left(\sigma_{ICA} / 0.2971\right)^{n} \quad (2.65)$$

$$K_{lapse} = \left[0.1156 \times BPR^{0.3379} - 0.004 \times BPR^{1.3379}\right] \times M_{cruise}^{-0.761} \quad (2.66)$$

where
σ_{ICA} is the density ratio at initial cruise altitude (ICA)
n = 0.75 for altitudes less than or equal to 36,089 ft (tropopause)
= 1.00 for altitudes above 36,089 ft

Note**:** If the buffet limited initial cruise altitude is less than 33,000 ft, then set the initial cruise altitude as optimum cruise altitude at top of climb + 2000 ft. This is to ensure adequate thrust available at top of climb for operational flexibility.

The thrust required per engine can now easily be calculated as follows:

$$T_{req} = T / W \times MTOW \div NOE \quad (2.67)$$

where
T/W is the thrust-to-weight ratio of the sizing scenario
MTOW is the maximum take of weight (lb)
NOE is the number of engines

Fuel Volume

The key characteristics of our aircraft such as the design weights, aerodynamic efficiency, high-lift architecture, and engine size are now known. The final step in our sizing process will be to make sure that the wing has the capacity to carry the required fuel and also have room for growth for the stretch model in our family.

Before we venture into the required fuel volume calculation (which can drive the wing area), it is a good idea to understand the reasoning behind the fuel volume requirement. Jet transports are typically designed to have some flexibility in their payload and range capability. The payload-range diagram shown in Figure 2.12 is a good tool in understanding the operational envelope for any aircraft. In a single snapshot, we can get a good idea of the design intent of any transport airplane.

FIGURE 2.12 The payload-range diagram.

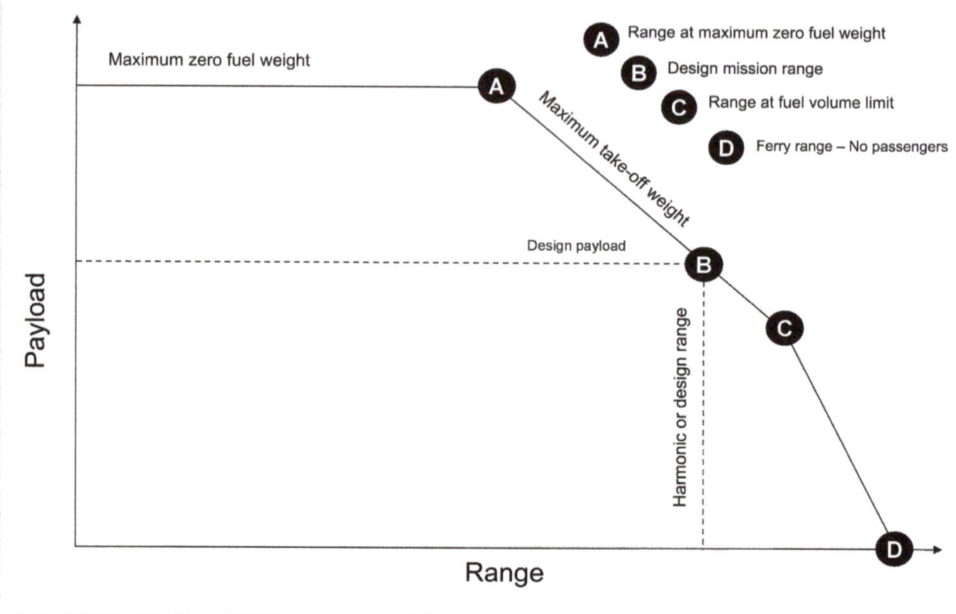

The line connecting point (A) to point (C) reflects the payload-range capability of the aircraft at MTOW.

Point (A) defines the maximum range of the aircraft at the maximum payload capability (MZFW) while flying at the MTOW. If the aircraft was operating below its MTOW but was still carrying its maximum payload, then it would lie on the horizontal line that connects point (A) to the Y-axis of the chart.

Point (B) on the chart is the design point that is set by our requirements to fly a certain payload over a certain range at the MTOW. This is also commonly referred to as the harmonic range of the aircraft. If we increase payload, we will have to de-fuel the aircraft to maintain its MTOW. If we reduced the payload of the aircraft, we could still operate the airplane at its MTOW by adding fuel in the wings, up to the point where the wing runs out of fuel volume. This is point (C) on the payload range chart.

The fuel volume of the wing should meet the need for flexibility to increase the aircraft range from point (B) to point (C) at MTOW. Different manufacturers have different methods that help determine the fuel volume to arrive at point (C); however, they all seem to collapse to an average delta of 5% over the mission fuel volume. Hence, the fuel volume of the wing is the fuel required to meet the design mission plus an additional 5%. It is important to note that too much fuel volume could drive the wing to be larger than necessary with the corresponding weight and drag penalties. On the flip side, a point designed fuel volume limited wing can limit the growth potential of an aircraft family.

A good case in point is a comparison between the Airbus A320 family and the Boeing 737 Next Generation family. The 737 has about a 50 ft^2 larger wing area than the A320.

Correspondingly, the 737 has more fuel volume; as a result, the 737NG family has been able to increase its range capability far more than the A320 family. The limitations of the A320's wing area is evident because of the use of body tanks to expand the aircraft's range. In the case of the A320 family, the choice of going with single-slotted trailing-edge flaps for the high-lift system also necessitated larger flap chords, which resulted in a smaller wing box and reduced fuel volume. A slightly larger wing on the A320 may have provided a better balance between field performance and range growth. Nevertheless, the A320 is a superb aircraft and offers very good operating economics. Recent advances in engine technology such as the geared turbofan have improved fuel burn, thereby reducing the amount of fuel required for a given mission. This is not an insignificant point, as evidenced by the A320 NEO family, where the improved engine and drag reduction technologies have allowed the wing–engine combination to be better matched. Although this was likely not in the initial sizing plan, it has also allowed Airbus to compete via an overall lower build cost that the smaller wing enables.

Note**:** Depending upon the mission, the wing area may be sized by approach speed, take-off, climb, and/or fuel volume requirements. In fact, with more sophisticated sizing tools which can size multiple aircraft families at a time, the wing area may be set by neither of the previous criteria for some family members. In some cases, the wing area may be set by an economic objective such as direct operating cost (DOC), which may seek to achieve the best balance between engine size and wing area without being pushed to the limits of the approach speed, climb, and take-off requirements. However; as mentioned earlier, this must be weighed against the build cost.

Jet fuel has a density of 6.7 lb/gal in Western countries, but in Asia, the fuel has a typical density of 6.5 lb/gal. The lower bound for jet fuel (Jet A) densities is about 6.3 lb/gal. The median value of 6.5 lb/gal is a good number to use to drive the wing fuel volume requirement. Note that sustainable aviation fuels, which are not yet prevalent, can have even lower densities, and these impacts should be considered when designing a future jet transport.

Fuel volume available in the wing is calculated as follows:

$$W_{fv} = \left\{ A \times \left(S_{wrp}^{\ 1.5} \times t / c_{outbd} \right) \div \text{SQRT}\ [AR] \right\} + B \tag{2.68}$$

where
W_{fv} is the wing fuel volume (gal)
$S_{wrp} = S_{ref}/\text{Cosine (dihedral angle)}$ (ft^2)

Here, the wing reference area is without the wing tip device (if present)
AR is the geometric wing aspect ratio without the wing tip device
t/c_{outbd} is the average outboard wing thickness-to-chord ratio (same value used to calculate the drag divergence Mach number)

A and B are constants that can be estimated from Figure 2.13

FIGURE 2.13 Wing fuel volume estimation.

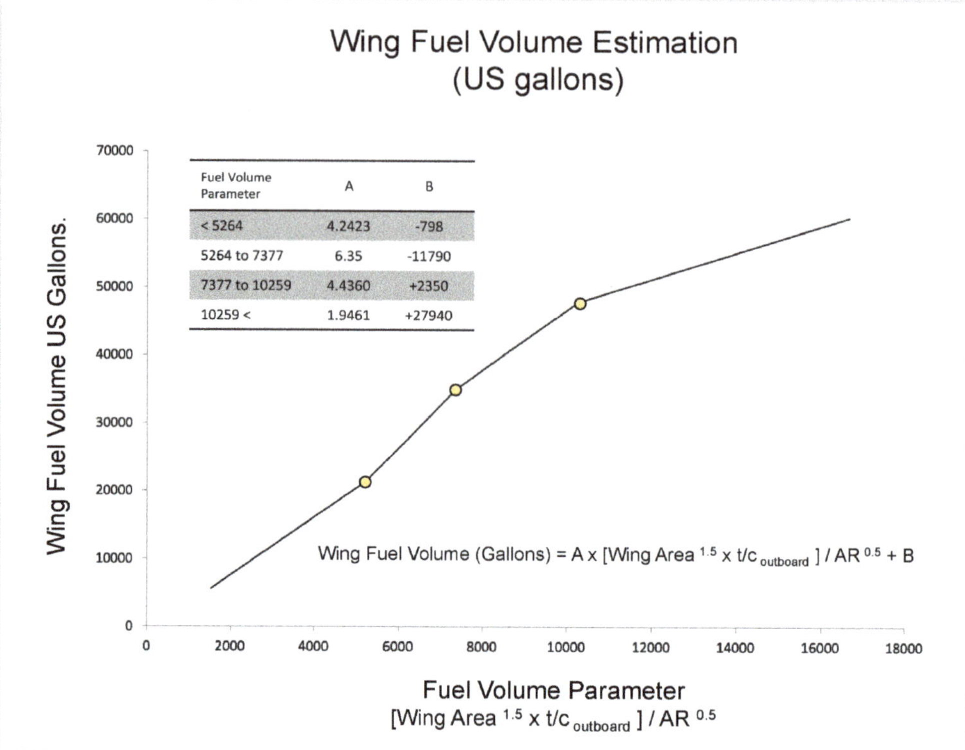

Limitations of the Sizing Method and Optimization

Before we move on to resizing our aircraft, it is prudent to understand the limitations of the method prescribed. The sizing method employed in this book is what is known as a "tops -down" method. In other words, this method is based upon parametric trends of take-off weight and OEWs and other high-level aircraft parameters. Tops down methods, in general, are powerful in that they minimize the source of errors at the early design stage, but this, in turn, is also one of their prime limitations. Bottoms -up methods, on the other hand, are those that depend upon weight buildups from the component level (wing structure, fuselage structure, systems, etc.), and they can offer insights into the effect of geometry on weight and aerodynamic efficiency.

A common notion is that bottoms -up approaches are more accurate because they tackle the physics of the problem from a component level; however, at the early design stage, these can also be misleading. In fact, most bottoms up approaches in the early design phase also rely extensively on parametric trends based on existing designs. These trends can vary depending upon the aircraft type, type of mission being flown, and design philosophy of the aircraft manufacturer. In his airplane design series, Dr. Roskam clearly defines several different methods in his weight analysis section so that the reader can choose the appropriate method depending upon the type of aircraft being analyzed. Increasing the number of individual component weight calculations can also increase the probability of error being introduced into the sizing process. In short, in the absence of detailed aircraft geometry and loads data, several different methods may be used as long as their limitations are well understood.

A good example of where the tops down method can break down is in the choice of wing aspect ratio or specifically wingspan. In our example, we chose a wing aspect ratio of 9.5, but we could have chosen a higher or lower aspect ratio, and the choice we made; while an educated guess is rather arbitrary. For example, if we chose an aspect ratio of 11, then the "tops down" method applied would capture the aerodynamic advantage of the higher aspect ratio wing but not the effect of the increase in wing weight. This, in turn, would indicate an exaggerated improvement in airplane fuel burn that could be misleading.

Before we draw up our airplane, we should confirm if we are in the ballpark of an optimum wing area and span (hence, aspect ratio) for our aircraft family. Furthermore, while on the topic of aspect ratio, it should be made clear that when we say aspect ratio, we mean the aerodynamic aspect ratio that is dependent on the effective aerodynamic wingspan.

A simple metric is available that can help us narrow down the aspect ratio of choice. This metric can be evaluated at the design range or at a different specified range. It is standard industry practice to design an aircraft to one range but actually assess its economic efficiency at a lower range. In the case of a short-range single-aisle aircraft, this economic mission range is typically set at 800 NMI. Hence, even though the Embraer 190, Airbus A320, and Boeing 737 are capable of ranges anywhere from 2200 to 3300 NMI, their economic efficiency will usually be evaluated by the industry at 800 NMI. Wide-body medium- and long-range aircraft are usually evaluated at 3000, 4000, and 6000 NMI, respectively. Airbus in practice prefers to use 4000 NMI for their medium-range economic mission, while Boeing usually opts to evaluate their economic performance at 3000 NMI.

There is a good reason why the industry has adopted this method for evaluating aircraft. A lot has to do with how these aircraft are operated. While aircraft such as the B787 and A350 are designed to fly 7500+ NMI, it is not unusual that they will fly shorter stage lengths during a given week. In fact, the average stage length for the above two aircraft can be around 4000 NMI (an 8.5 hr flight) depending on the airline operation. Hence, an aircraft should be flexible enough to provide an economic benefit to its operator over short, medium, and long ranges. This is a fairly profound requirement. It requires that an aircraft be both weight efficient and aerodynamically efficient. Weight efficiency goes a long way to improving an aircraft's fuel burn in all ranges of flight, while cruise aerodynamic efficiency has the effect of improving fuel burn in the longer ranges. However, it should be noted that short-range aircraft in particular are more dependent on weight efficiency over cruise aerodynamic efficiency. An added facet to the problem is that an improved overall aerodynamic efficiency leads to a lower fuel load for a given mission and a lower structural weight. Hence, airplane programs spend hundreds of millions of dollars in their quest to reduce both weight and drag through investments in advanced materials, propulsion and aerodynamic technologies.

Note: If we were evaluating an airplane with an economic mission range below 5000 NMI, then we would apply a range correction factor. This is because in most short- and medium-range missions, the range covered in climb and descent is becomes significant, and the block fuel sensitivity is increased. In such cases, the economic mission range would be **{ECON-MISSIONRANGE – 200} NMI**.

The efficiency metric is a rather simplistic formula (applicable to tops down methods), and it allows us to optimize the wing area and wingspan for a given economic mission. In order to accomplish this, we "resize" the airplane (multiple wing areas (and wingspan) and thrust) for the design mission range while evaluating its performance for a shorter economic mission range.

The optimization parameter or efficiency metric is a weight and fuel burn metric, which is given by the following equation:

$$W_{fm} = \text{Sqrt}\left\{BF_{\text{econ-mission}}^{2.1} + Wt_{\text{wing}}^{2} + Wt_{\text{propulsion system}}^{2}\right\} \tag{2.69}$$

where
W_{fm} is the weight and fuel burn metric = optimization parameter or efficiency metric
$BF_{econ\text{-}mission}$ is the block fuel for the economic mission (mission fuel – fuel reserves) (lb)
Wt_{wing} is the wing weight (lb)
$Wt_{propulsion}$ is the total propulsion weight (includes the weight of the strut, nacelles, and thrust reversers) (lb)

The wing weight can be obtained from the following equation:

$$Wt_{\text{wing}} = K_w \times b_s^{0.75}\left\{1 + \text{SQRT}\left[6.25/b_s\right]\right\} \times n_{\text{ult}}^{0.55}\left\{\left[b_s/t_r\right]/\left[\text{MZFW}/S\right]\right\}^{0.30} \times K_e \times K_c \tag{2.70}$$

where
K_w
= 0.0019 for a single-aisle aircraft
= 0.00164 for a twin-aisle aircraft
$b_s = [b_{effective}/\cos(\theta_{c/2})]$
$b_{effective}$ is the effective aerodynamic span (ft)
$\theta_{c/2}$ is the sweep at the half chord line of the wing
n_{ult} is the ultimate G load = 3.75 for transport aircraft
t_r is the maximum thickness at the root (side of body) (ft)
$= 2 \cdot t/c_{root} \times (S/b)(1 + \lambda)^{-1}$; here, b is the wingspan without any tip device and t/c_{root} is the maximum thickness-to-chord ratio at the root
MZFW is the maximum zero fuel weight (lb)
S_{ref} is the wing reference area (ft^2)
K_e
= 0.95 for engines mounted on the fuselage (e.g., MD-80 and Gulf Stream V)
= 0.83 for four-wing-mounted engines (e.g., B747, A340, and A380)
= 0.85 for tri-jets and twin-engine airplanes (e.g., B787, B777, and A330)
K_c
= 1 for aluminum wings (e.g., B777, B767, and A330)
$= 0.85 \times 0.998^{(EIS - 2008)}$ for composite
EIS is the entry into service date

The propulsion weight can be calculated as follows:

$$Wt_{\text{propulsion}} = 1.57 \times Wt_{\text{Dry engine}} \times N_e \tag{2.71}$$

where
N_e is the number of engines
$Wt_{Dry\ engine}$ is the dry engine weight without the nacelles, fuel system, etc. (lb)

$$Wt_{Dry\ engine} = 1.67 \times \left(F_n^{0.45} \times \text{Fan diameter}^{0.6}\right) / \text{TSFC}^{0.56} - K_{eng} \quad (2.72)$$

where
F_n is the sea level static thrust per engine (lb)
Fan diameter is the engine fan diameter (in.)
TSFC is the thrust-specific fuel consumption
K_{eng}
= 0 (for a direct drive engine)
= 11.2 × Fan diameter (for a geared turbo fan or GTF engine)

$$\text{Fan diameter} = 0.231 \times \left(F_n^{0.5} \times \text{BPR}^{0.25}\right) + 4.71 \quad (2.73)$$

A plot can now be created such that the optimization parameter is plotted versus the wing area for multiple family members. Wing area (and hence, wingspan at a constant aspect ratio) is varied to make certain that all the necessary constraints (approach speed, TOFL, climb, fuel volume, etc.) are met.

FIGURE 2.14 Aircraft family wing area selection.

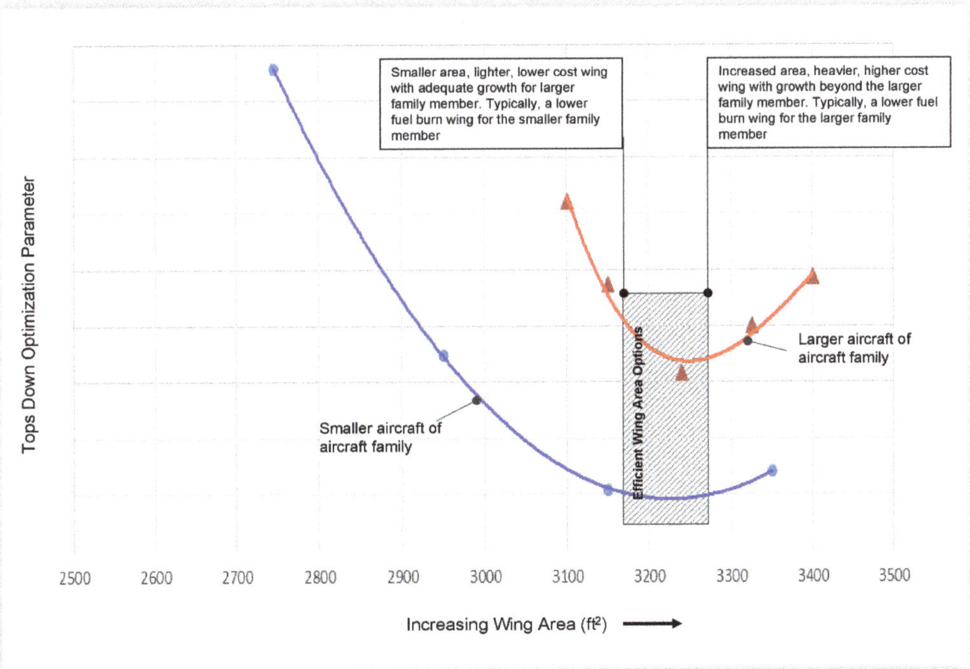

Also note that the optimum wing area and corresponding wingspan can vary with different economic mission stage lengths, even though the design mission remains unchanged. For example, the design mission range can be 7300 NMI but optimizing for 3000 NMI versus a 6000 NMI economic mission range can lead to a different wing area and wingspan solution. There is also the consideration of which category of gates the airplane can operationally speaking fit into. The distinct wingspan limits are listed in Table 2.9.

TABLE 2.9 Wingspan limits and ICAO airport codes.

Wingspan limits (ft)	ICAO—airport reference code
Wingspan < 49	A
49 ≤ wingspan < 79	B
79 ≤ wingspan < 118	C
118 ≤ wingspan < 171	D
171 ≤ wingspan < 213.25	E
213.25 ≤ wingspan < 262.5	F

Practically speaking, an airplane will be used differently by different operators, so the term optimization can be a bit of a misnomer. In reality, there is no absolute correct philosophy in picking a wing area and the corresponding wingspan. This is true especially when one is designing a family of aircraft. It is up to the airplane designer to ultimately choose how they want to bias the wing design (for weight or drag). One of the most important considerations in choosing wingspan and wing area is the integration challenges associated with drawing up a wing planform, that is, making sure that there is adequate room for systems, flaps, ailerons, etc., while preserving wing box area to meet the fuel volume requirements. The wing box plays a significant role in counteracting high air loads that induce wing bending and torsion. Integrating the wing main landing gear to make sure it is in the right position for airplane balance and that it fits inside the wheel well is also a key consideration in wing planform layout. It is this author's experience that to achieve a well-integrated wing with a quarter chord wing sweep of more than 30°, the basic wing aspect ratio (wing without the wing tip device) of approximately nine will probably be the upper limit. Adding advances in high lift design such as spoiler droop usually take up precious systems volume and can reduce the integrated basic wing aspect ratio by another half unit. This is especially true of airplanes with wing mounted gear only. For airplanes with wing and body mounted gear (A380, B747, A340-600), an aspect ratio which is slightly higher is possible, if that is what is desired. This is because of the flexibility afforded by the body gear to locate the effective main landing gear (wing plus body gear) geometric center to a favorable location with respect to the mean aerodynamic chord. More on this will be discussed when we get into wing planform design.

In practice, barring the academic exercise to seek out an optimum performing airplane, there are also considerations such as infrastructure (factory floor space, autoclave sizes, etc.), and transportation logistics which deal with oversized airplane parts, such as wings and fuselages. In fact, in a world which is shrinking rapidly and with the increased trend toward outsourcing, the logistics of designing and building a large jetliner have become ever more challenging.

On the Boeing 787, a special Large Cargo Freighter was designed (A modified 747) to bridge the worldwide supply chain. Care was taken in the design of the 787-wing planform to make sure that a pair of wing box structures could be transported in a single operation. In the case of the A380, the governing politics of the Airbus partner countries (France, Germany, England, and Spain) weighed in on the manufacturing plan that was to be utilized. The sheer size of the aircraft necessitated water and road routes that took years of planning.

FIGURE 2.15 The Large Cargo Freighter designed for the B787 Production System.

Thiago B. Trevisan/Shutterstock.com.

FIGURE 2.16 A380 components being transported by ship.

A Photon Bounce Creation/Shutterstock.com.

Operating Cost

There are several aircraft manufacturing companies around the world that cumulatively manufacture approximately 20+ different types of commercial jet transports. As a result, there must be a standardized method of comparing the economic benefits of these aircraft under a standard defined set of conditions. This is a benefit to both the airline customer, who must make a selection from various different types of aircraft before committing an obscene amount of monetary investment, and the manufacturer, who literally bets the farm every time a new aircraft development venture is undertaken.

It is important to the aircraft manufacturer and the aircraft operator that the technical performance and economic benefits of each aircraft design are well understood over the lifetime of the product before committing such vast sums of money. Aviation history is littered with technically viable and even superior aircraft, which failed miserably because the manufacturer misunderstood or simply ignored the economic demands of the market. The Convair 880 and 990 series of aircraft and the Concorde are just a few examples of what can happen when the technical and economic goals of an aircraft program are not well aligned. It is for this very reason that The Boeing Company abandoned the Sonic Cruiser (due to rising fuel costs) and embraced the B787, which has also fulfilled its promise to be both an environmentally friendly and economically superior aircraft.

The economic benefits of a particular aircraft design are based on several key characteristics, which include fuel burn, number of flights per year, number of cockpit and cabin crew, cost of acquisition of the airframe and engines, maintenance costs, insurance, and the residual value of the product. A well-rounded design is one which is designed from the outset for low fuel burn, low maintenance, and high residual value. Maintenance costs and residual values are usually not well understood by most aircraft engineering departments and are also a lot harder to quantify early in product development. Hence, it is not uncommon to find a lot of effort being invested into things that are better understood such as weight and drag which impact fuel burn. The Boeing 787 has consciously attacked maintenance costs by adopting composite materials (to address airframe corrosion) along with a deliberate approach to design standardization. This, in turn, has had an enormous impact on residual value. The more standardization the aircraft has, the easier it is to move the aircraft from airline to airline without incurring large costs for refitting interiors, etc. In fact, the strong drive within the Boeing Company toward a composite airframe was motivated as much by maintenance as it was fuel burn. The design philosophy of the Boeing 787 ensures low fatigue and a virtually corrosion-free environment. This has a significant impact in the time elapsed between major aircraft checks and also the number of parts to be replaced during the check itself. Design methods such as using impact resistant materials applied to the door surround structures also ensure minimal downtime due to damage caused by jet ways and other ground support equipment (GSE).

Other unusual steps taken by Boeing included lowering the cabin altitude from the conventional 8000 to 6000 ft. This combined with larger windows was aimed at enhancing the travel experience of the general public by providing them with a more comfortable environment and re-connecting them with the magic of flight.

A good aircraft designer should be mindful of all design considerations that can enhance the product and avoid falling into the trap of being too focused only on performance-driven parameters such as L/D or TSFC. For example, on short-range aircraft such as regional jets and smaller airliners such as the 737 and A320, the aircrafts unit recurring cost and maintenance cost are as much a driver as fuel burn. In such aircraft, it may be appropriate to seek out a better balance between low cost to build, maintenance cost, and fuel burn. Engines designed for improved fuel burn sometimes achieve this by going to larger fan diameters, bypass ratios, and increased complexity and weight. Complexity can hurt maintenance costs, while higher bypass ratio engines could actually drive aircraft weight up and hurt the overall economics. The last statement, although generally applicable to all aircraft, is particularly true for shorter range aircraft, where weight is a key driver of aircraft economic performance. Geared fan technology seems to be bucking this trend by providing a larger fan, which is mated to a lighter core, thereby making the entire package more efficient.

Understanding the impacts of these key drivers for an economically successful product, we can define some specific metrics that make up the total operating expenses for an airline. The total operating expenses can be broken down into the following:

- Airplane price
- Airplane utilization per year
- Flight crew expenses
- Fuel and oil expenses
- Insurance expenses
- Other flying operation expenses
- Passenger service expenses
- Flight equipment maintenance expenses
- Flight equipment depreciation and amortization expenses
- Landing fees

Airplane Price

The first thing to establish is the approximate price for a given aircraft model. Several sources exist on the Internet (The Boeing website is a good source), which have information on existing list prices. Simple curve fits can be derived which relate range, payload, and fuel burn to aircraft price. The rationale here is to reasonably predict the price of an aircraft based on some value-added metrics such as range, payload, and fuel burn, all of which can drive aircraft price. The numbers given are in line with 2017 prices, and they can be adjusted for future products based on an average inflation rate of 1.1%. In other words, a factor of $1.011^{(\text{YEVA}-2017)}$ can be multiplied by the derived values below that are in 2017 dollars. Here, the term YEVA is simply the year in which the relevant aircraft models are being evaluated and not necessarily the year the model enters into service.

$$\text{Single-Aisle airplane price} = \{0.001527 \times (\text{Productivity}) + 45.5\} \times 1.011^{(\text{YEVA}-2017)} \tag{2.74}$$

$$\text{Twin-Aisle airplane price} = \{0.00194 \times (\text{Productivity}) + 88.5\} \times 1.011^{(\text{YEVA}-2017)} \tag{2.75}$$

$$\text{Productivity} = (\text{Seats} \times \text{Design range}^{0.85}) \div (\text{Block fuel per seat for design mission})^{0.23} \tag{2.76}$$

FIGURE 2.17 Aircraft price estimation.

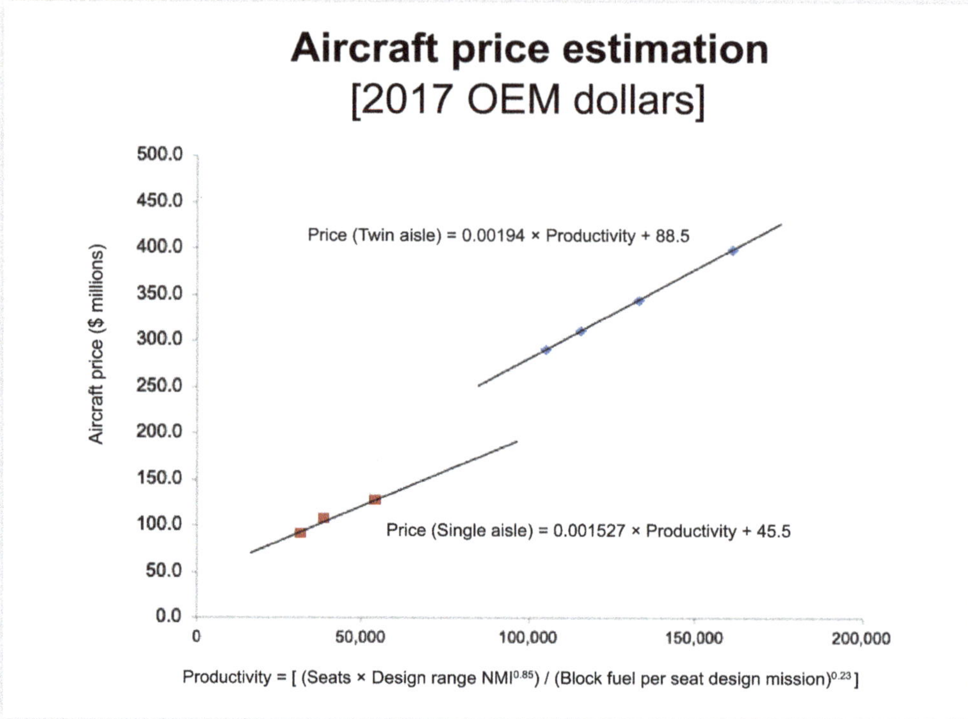

The productivity index was developed using existing airplane price [67, 68] and airplane performance data [16, 17].

For wide-body short- to medium-range airplanes with design ranges of up to 5000 NMI (productivity index range 75,000 to 95,000), pricing can become a challenge as these products straddle efficient and lower priced single-aisle aircraft on the lower end, and more capable wide-body aircraft such as the Boeing 767-300 on the upper end of the segment.

Utilization

Utilization is the estimated number of flights the aircraft will fly in one year. Utilization is usually calculated based on the economic mission range and not the airplane design range.

$$\text{Utilization} = \left\{680 + \left(0.0004 \times \text{Range}^2 + 6.5 \times \text{Range} + 1510\right) / \left(\text{BH} + 0.5\right)\right\} / \text{BH} \qquad (2.77)$$

Range is the economic mission range
= 500 NMI to 1000 NMI for short-range airplanes
= 3000 to 6000 NMI for long-range airplanes

$$\text{BH} = \text{Block hours} = \left[\left(\text{Range NMI} - 200\ \text{NMI}\right) / \text{Cruise speed knots}\right] + 1 \qquad (2.78)$$

Flight Crew Expenses

$$\text{Flight crew expenses} = \text{AF}\left\{\text{K}\left(\text{MTOW}\right)^{0.4}\right\}\left(\text{BH}\right)\$/\text{trip} \tag{2.79}$$

where
AF is the airline factor
= 0.8 average
= 0.63 low-cost airline
= 1.0 large efficient carrier

For most cases, use the average value unless the cost model of the airline is better understood.

Fuel and Oil Expenses

The cost of fuel is no doubt a variable and is difficult to predict from year to year. Various world events can have a profound impact on fuel prices, but over the long term, the oil prices typically adjust and remain relatively stable. In other words, while it can be expected that there will be market forces that cause spikes from time to time, it is a reasonable assumption that the global economics of our planet will ensure that no extreme cost is sustained for a long time. For example, the oil conglomerates are acutely aware that if they decide to raise oil prices beyond a certain threshold, the market will push for alternate energy sources, which would ultimately reduce the demand for their product. Competition from alternate energy sources combined with increasing pressure for a greener environment is the "check and balance" that will keep the price of oil within acceptable limits. Of course, one can only hope that circumstances demand, rational thought will prevail.

Figure 2.18 shows the changing price of aviation fuel with time and can be used as a guideline when doing a DOC calculation.

FIGURE 2.18 Fuel costs with time in US dollars.

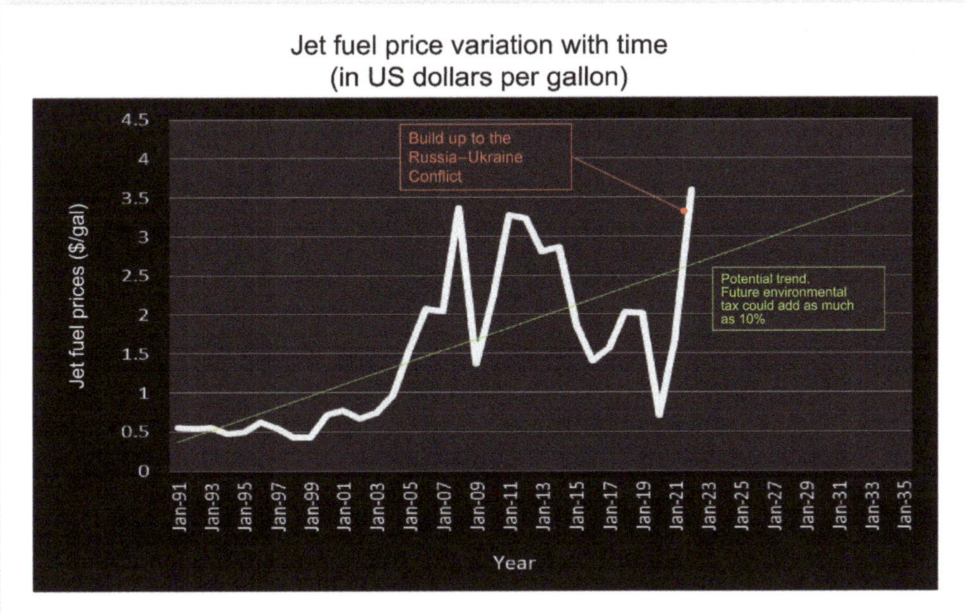

The average cost of fuel from 1986 through 2003 has been around 63 US cents per gallon. The distinct spike in fuel cost just after 2003 is due to the instability in the Middle East due to the second Persian Gulf War and the uncertainty it represented. In March of 2007, the price of fuel was approximately 180 US cents per gallon. If one uses the historic trends to predict the average cost of fuel over the next decade, one may conclude that fuel prices may average out to about 300 cents per gallon. This is, of course, just a guesstimate. No one can reasonably know the number given the vast uncertainty in the energy situation, but this number, it is hoped, will capture the order of magnitude of fuel impact on operating cost. Furthermore, given the growing pressure to keep carbon emissions low along with reducing noise and other environmental impacts, it is reasonable to expect some sort of environmental tax. Without giving it too much thought and learning what we can from existing press, we will assume this value to be approximately a 10% premium added to the cost of fuel.

$$\text{Fuel expenses} = \text{Block fuel} / 6.5 \times \text{FP} \ \$ / \text{trip} \tag{2.80}$$

From a stand point of calculating fuel expenses, our choice to use 6.5 lb/gal was rather arbitrary, and we could have used 6.7 lb/gal. What is important is that for comparison purposes, you use the same value between different aircraft models whether it is Boeing, Airbus, or any other manufacturer.

Insurance Expenses

The insurance expenses here are related to the hull insurance carried by an airline per aircraft per year and is based on the aircraft purchase price. The rates are intended to reflect 2006 values, but, of course, in reality, there are several factors that would play into the actual number.

$$\text{Insurance} = 0.00825 \times \text{Purchase price} \tag{2.81}$$

On a per trip basis, the cost can be expressed as Insurance ÷ Utilization $/trip.

Other Flying Operation Expenses

These expenses are related to several miscellaneous items that have to do with flight operation-related expenses.

$$\text{Other flying operation expenses} = 0.04 \times \left(\text{Flt crew} + \text{Fuel} + \text{Insurance}\right) \$ / \text{trip} \tag{2.82}$$

Passenger Service Expenses (PSE)

These expenses are related to flight attendant pay in 2006 dollars and can be expressed as follows:

$$PSE = 1.6 \times (67{,}600) \times (\text{Number of flight attendants}) \div \text{Utilization } \$/\text{trip} \tag{2.83}$$

67,600 = average flight attendant expenses in the US in 2006 dollars (assumed to include benefits).

$$\text{Number of flight attendants} = \text{No.of passengers} \div K_f$$

where
K_f = 33 short-range aircraft flying ≤ 3500 NMI
= 23 long-range aircraft flying > 3500 NMI

Flight Equipment Maintenance Expenses

$$\text{Airframe maintenance} = K\left\{(\text{Ref W})^{0.72118}(\text{FH})^{0.46050}(\text{DP})^{0.32062}\left[1+(\text{In house AF}/\text{Total AF})\right]^{-0.43177}\right\}\$/\text{Year} \tag{2.84}$$

$$K = 2.0 \times MF \times CF \tag{2.85}$$

MF is the Aircraft Model Factor = $0.9985^{(EIS - 1994)} \times 0.35$. The factor determines how new the aircraft model is, and it is an attempt to capture airframe maintenance improvement that is incorporated into new aircraft as they are introduced into the system.

CF is the airline cost factor
= 1.0 average
= 0.8339 low-cost airline
= 1.3019 major well-run carrier

$$\text{Ref W} = \text{Reference airframe weight} = \text{OEW} \times K - \text{Engine dry weight}$$

Here, K has a value as follows:
= 0.935 single-aisle aircraft (A320, B737)
= 0.93 twin-aisle double-deck aircraft (B747-8, A380)
= 0.92 twin-aisle single-deck aircraft (B777, A350)

Engine dry weight is calculated from Equation (2.72).

Note: In an aircraft family, typically, the engine is sized by the larger family member and derated for the smaller initial mode. The thrust used in determining the engine size will therefore be that of the larger/more capable mode in the aircraft family.

$$\text{FH} = \text{Flight hours per year} = \{(\text{BH} - 0.5) \times \text{Utilization}\} \tag{2.86}$$

DP = departures per year = utilization = 645 departures.

(In house AF/Total AF) = Ratio of in-house airframe maintenance to total airframe maintenance.

If all of the maintenance is done in house, then this ratio is 1.

$$\text{Airframe maintenance} \div \text{Utilization gives the value in \$ / trip} \tag{2.87}$$

$$\text{Engine maintenance} = \text{K}\{(\text{Thrust})^{0.8965}(\text{NE})^{0.9234}(\text{FH})^{0.15344}(\text{DP})^{0.37535} \left[1 + (\text{Outside engineering / Total engineering})\right]^{-0.34704}\} \tag{2.88}$$

Here,

$$\text{K} = 2.05 \times \text{MFE} \times \text{CF} \tag{2.89}$$

where MFE is the model factor for engine.

$$\text{MFE} = 0.9985^{(\text{EIS}-1994)} \times 0.35 \times \left[(\text{NE} + 2)/4\right]^{0.35} \tag{2.90}$$

Thrust = thrust per engine = actual thrust required by the aircraft to perform its mission. This shows the importance of derating (or using less thrust when required), which has a direct impact on engine wear and tear. Typically, airframe–engine combinations across the family are not derated beyond 10% when sized.

NE is the number of engines. This is one of the major reasons why twin-engine airplanes are more cost efficient than quads or tri-jets. The more engines you have, the higher are your maintenance costs. Important strides made in engine reliability have allowed this trend to continue.

The only situation where having more than two engines might make operational sense is in high-altitude and hot-temperature airfield operations. In such circumstances, a twin-engine airplane may be more severely limited in the amount of payload it can fly out of an airport for a given range. A good case in point is a particular carrier's choice for the A340-600 over the 777-300ER. Here, the airline was seeking an airplane that could fly certain long routes out of a high and hot airfield, and the 777-300ER was not able to do that mission with a full passenger payload plus cargo. Hence, even though the 777-300ER was the airplane of choice due to its better economics, that particular airline mission constraint cost Boeing the campaign.

FH is the flight hours per year.

$$FH = \{(BH - 0.5) \times \text{Utilization}\} \tag{2.91}$$

DP is the departures per year = utilization = 645 departures.

Outside engineering/Total engineering = Ratio of engine maintenance done outside to total engine maintenance.

If all of the maintenance is contracted to an outside maintenance shop, then this ratio is 1. This is the number we will assume for all our calculations, but an airline using these formulas can input their assessment value depending upon the level of work they subcontract to an outside engineering shop. Note that most airlines subcontract most if not all of their engine maintenance to outside specialized shops.

Engine maintenance is in $/year.
Engine maintenance ÷ Utilization gives us $/trip.

Flight Equipment Depreciation and Amortization Expenses

This expense category has to do with depreciation of the aircraft as an asset, and the residual value an airline may hope to recoup after the aircraft has spent 20 years in service. The residual value assumed here is 15% of the value of the aircraft purchase price, which is a typical value. However, several market factors can play in to improve or reduce the residual value of the aircraft. For example, the 787 with its greater standardization of equipment between airline customers and better corrosion resistance due to composites will have a supposed premium on residual value. This is an enticing prospect for would-be financiers (banks) who are seeking to get a high return on their investment over the long term. The recent demand for wide-body aircraft in the market place is also serving to keep residual values of such aircraft high. This in spite of the increased fuel prices which are typically less favorable to older aircraft.

$$\text{Depreciation and amortization expenses} = APP(1 - RV) \div DP \tag{2.92}$$

where
APP is the aircraft purchase price
RV is the residual value
= 0.15 for conventional aluminum aircraft
= 0.20 for composite aircraft
DP = depreciation period
= 20 yrs
Depreciation and amortization expenses is in $/year.
Depreciation and amortization expenses ÷ Utilization gives us $/trip.

Landing Fees

Landing fees are those fees that an airport charges an airline based on the MLW of the aircraft.

$$\text{Landing fees} = 0.00147 \times \text{RF} \times \text{MLW} \tag{2.93}$$

RF is the route factor that tries to capture the average charge by region of the world
= 1 US domestic
= 2.36 Atlantic routes to Europe
= 4.28 Pacific routes to Asian countries (include over flight of Russia and China)
= 1.64 routes to Latin America

Of course, as matter of good design practice, when comparing any two aircraft, it is probably good to have a consistent number for the comparison. The only place this may make a difference is if one particular aircraft, for example, does not have the design range to do a direct flight from, say, the US to Asia and may need to stop over in Europe before continuing on to Asia. In this case, it would be appropriate to charge the aircraft a landing fee charge for that stop in Europe plus the final landing charge in Asia.

Remember, the purpose of calculating the operating expenses is not necessarily to obtain the absolute value for a given aircraft, but more so to have a consistent means of comparing two aircraft in a design study or competition.

Landing fees are in \$/trip.

Total Aircraft Operating Cost

The total aircraft operating cost (TAOC) can then be determined by adding up the individual components:

$$\begin{aligned}
&\text{Total aircraft operating cost} \\
&\quad = \{\text{Flight crew expenses} + \text{Fuel and oil expenses} + \text{Insurance expenses} \\
&\quad + \text{Other flying operation expenses} + \text{Passenger service expenses} \\
&\quad + \text{Flight equipment maintenance expenses} \\
&\quad + \text{Flight equipment depreciation and amortization expenses} + \text{Landing fees}\}
\end{aligned} \tag{2.94}$$

TAOC are in \$/trip.

The total aircraft operating cost can also be calculated in per seat per nautical mile metrics.

Typically, this can be compared to other aircraft in a category by placing the trip costs and per seat costs on a graph. The graph in Figure 2.19 is typical of the type of charts found in industry, and it can serve multiple purposes. It is commonly used to compare aircraft families of different sizes to evaluate fuel burn and separately operating cost, which, in the case of TAOC, includes the price of the aircraft. When starting out on a new project, charts like this can also be used to set the economic targets and then used to drive technologies to achieve those targets. In this respect, they can be very useful.

FIGURE 2.19 Total airplane operating cost (including ownership).

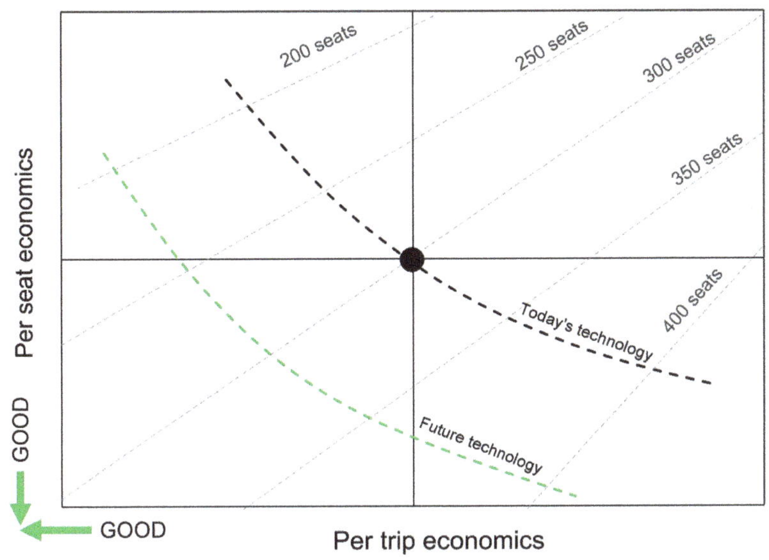

The x-axis represents the metric per trip, while the y-axis represents the metric per seat. This chart also does not capture the effect of design range, and typically, an aircraft that flies farther (utilizing the same level of technology) will have higher costs than an aircraft that has less design range. Hence, charts like this need to be put in the right context and viewed holistically along with other relevant data, such as payload-range capability and field-length range capability. The latter being important because it demonstrates how much revenue-generating payload (passenger + cargo) a transport can carry out of a given airfield to a destination.

One way to look at this chart is to view the x-axis as a risk indicator. For example, a large aircraft such as the Boeing 747-8 or A380-800, both can show an improvement in per seat economics (bottom right quadrant) over a medium-capacity aircraft, but it will have higher absolute operating economics. In other words, airlines that operate the 747-8 and the A380-800 have to make sure they have high load factors (high number of passengers) in order to realize the full per seat economic potential. Once the airplane is designed and it is in revenue service, the cost to operate a particular flight (crew costs for instance) is fairly constant, and the revenue generated depends upon the number of tickets sold for that flight. Hence, if an airline feels they have the passenger traffic to sustain the cost of operating a large aircraft, then it makes sense to operate a large aircraft with 400 to 500 seats, instead of two smaller aircraft with 250 seats on high-traffic routes. This is particularly true of routes to and from major hubs like London Heathrow and New York, John F Kennedy (to name a few), which are congested and becoming slot constrained.

You can also see from the chart in Figure 2.19 that advances in technology can shift the economic horizon down and to the left to the lower left quadrant. This is where new aircraft economics should go directionally, when being compared to a previous generation or existing generation transports. The use of composites along with advanced engine and systems technologies has now made it possible to achieve both superior seat-mile economics and also lower the absolute cost per trip. In essence, they reduce the risk to an airline which may have routes that are not high-traffic routes.

The industry has spoken, and in 2019, Airbus cancelled the A380 program due to poor sales. No doubt the European manufacturer needed a competitor in the large airplane market, but whether they went too large in terms of aircraft size is now no longer debatable—They overshot on the size of the A380. It is this author's opinion that perhaps Airbus would have been better of taking the 747 head on with a 400-seat aircraft, which could then be stretched to 450 seats, if necessary. The current A380-800 wing is sized by the A380-900 (630+ seats), and the aircraft pays a hefty performance penalty in both weight and drag for the larger wing area.

One could speculate that with all the talk in the mid-1990s about large airplanes, Airbus realized that if it was to build a large aircraft, it would have to be a family in the 550-to 650-seat category. The Boeing 747-500 and 747-600 family being studied at the time could not grow much beyond 550 seats. To compete directly in this category (450 to 550 seats) with Boeing would have given the American plane maker an advantage of a lower cost to market derivative aircraft. The derivative approach would also give Boeing the potential advantage for an aircraft with a very competitive OEW and total investment. This essentially may have dictated that Airbus would require a 550-to 650-seat aircraft family to have a viable program for the long term.

At the time, Airbus was also developing the A340-500/600 family to get into the 300+ seat market. With the A340-500/600 Airbus' competitive offering in this seat category was deficient and limited in growth, given that it was an 8-abreast cabin. The twin-engine economics of the Boeing 777 ultimately led to the demise of the A340 family. Airbus finally responded to this market need with the A350-XWB family, which is doing well in the market space.

Getting back to the late 1990s, meanwhile, Boeing continued to evaluate the market and discovered that with the right mix of advanced technologies, a smaller aircraft could have break through economics, which would beat the seat-mile costs of a conventional technology large aircraft. This is perhaps the key element that Airbus may have missed in their analysis.

Having said that, the author has flown on the A380, and it is liked by both passengers and crew. It also provides the airline operator the opportunity to have increased revenues if they can attract the high-yield business traveler and fill the airplane. Revenue potential, which the A380 has in spades, is not typically calculated in the early phase of the design process, but it should not be overlooked.

Configuring the Airplane

This is by far the most interesting part of the airplane design process. While aircraft sizing is the art and science of guesstimating how a future project aircraft should perform, configuring the aircraft is the art of making sure that the as-drawn aircraft meets the predictions of the sizing process.

The general sequence of laying out an airplane is outlined below and will be described in the chapters to come.

- Fuselage cross-sectional design
- Fuselage layout
- Propulsion
- Wing design and landing gear layout
- Horizontal tail
- Vertical tail
- Integrating the airplane

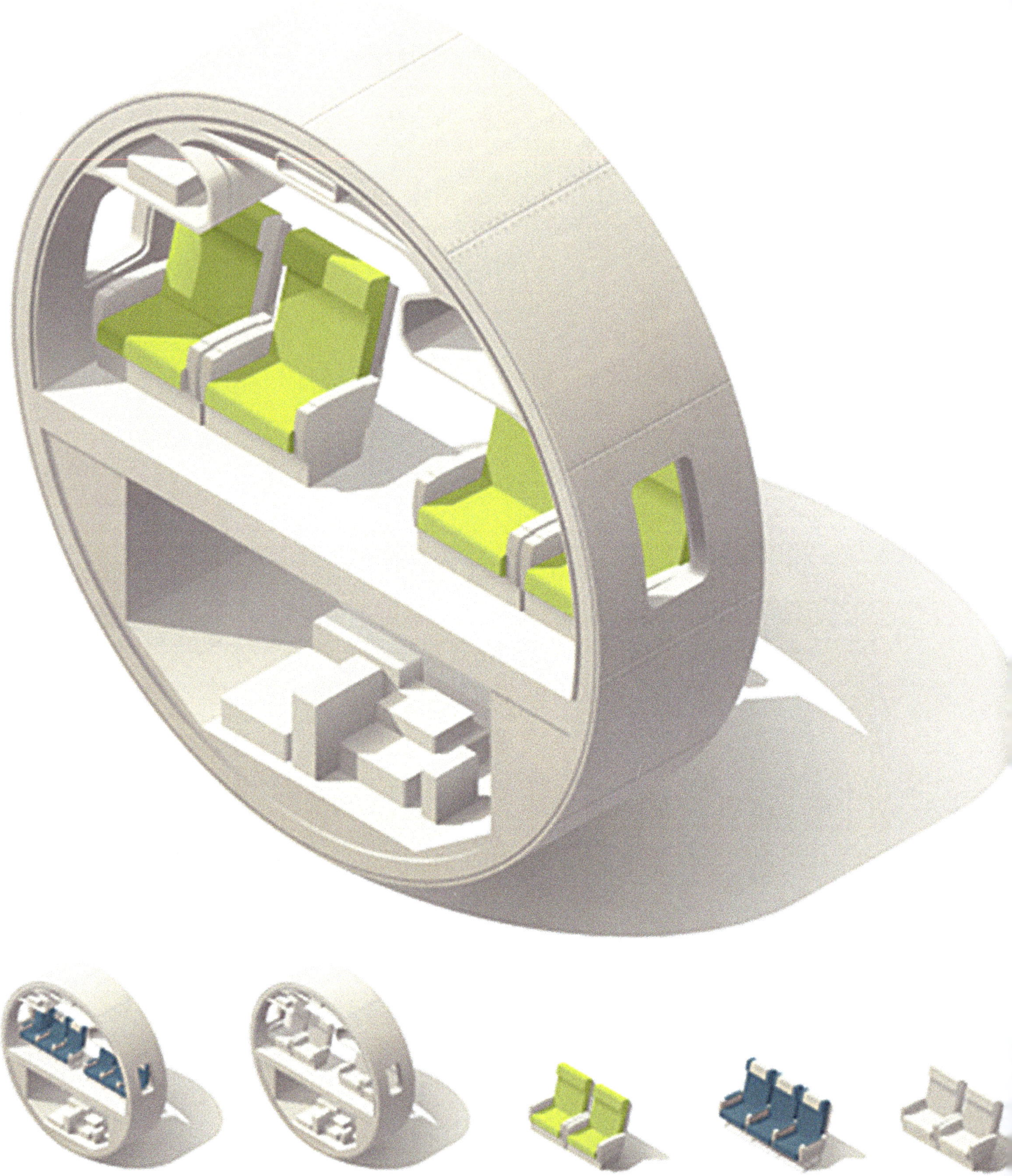

Tele52/Shutterstc

3

Fuselage Cross-Section Design

Draw out a horizontal floor

$$\text{Length} = 0.5 \times \text{max.diameter}\,(\text{in.}) \tag{3.1}$$

$$\text{Depth} = \text{Economy seats abreast} \times 0.9\,(\text{in.}) + 0.5\,(\text{in.}) \tag{3.2}$$

Begin placing the seat combinations starting from the outside edge of the floor toward the center of the cross section. It is good drawing practice to do one-half of the cross section and then mirror the image across the centerline. This assures symmetry. Place the seats such that the bottom of the seat cushion is 9.5 in. off the top of the floor. Since we will be drawing a nine abreast airplane in a 3-3-3 format, each side will have 4.5 seats such that there will be a triple seat separated by an aisle followed by half a triple as shown in the figure. The minimum aisle width acceptable for a galley cart is 16.5 in., but it is good design practice to not go below an 18-in. aisle for practical reasons.

FIGURE 3.1 Layout of floors and seats.

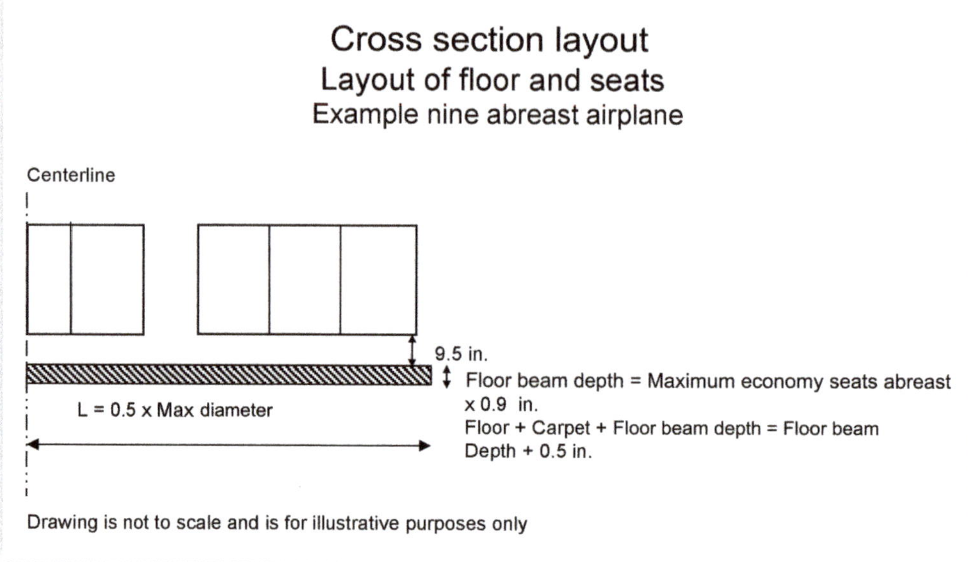

FIGURE 3.2 Clearance values for a seated person.

Cross section layout

Clearance values for a seated person

Example nine abreast airplane

Note: All clearance values are radii and in inches

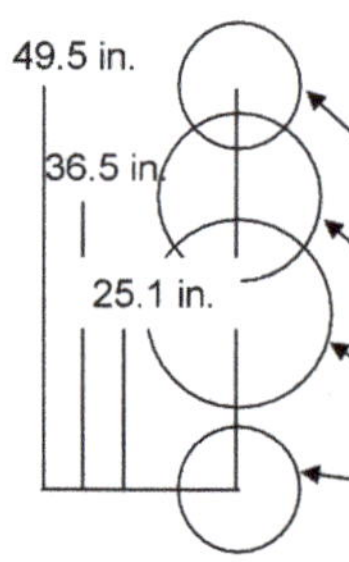

18 in. Seat bottom	17.2 in. Seat bottom	17.5 in. Seat bottom	18.5 in. Seat bottom	19.0 in. Seat bottom	20.0 in. Seat bottom	21.0 in. Seat bottom	Minimum recommended
7.3	6.7	6.9	7.7	8.1	8.8	9.6	6.0—Long range 5.5—Short range
10.6	10	10.2	11	11.4	12.1	12.9	9.5—Tourist class
11.9	11.3	11.5	12.3	12.7	13.4	14.2	11—Tourist class
9	8.4	8.6	11.4	9.4	10.5	11.3	9—Long range 6.5—Short range

The vertical heights are derived from Reference 23, 25, 28, and 29. The shoulder height and eye height being representative of a 95th percentile male.

The dimensions of the circle are guidelines based on information found on the Aircraft Design website of Stanford University and Reference [21]. The cross section shown in the reference document is a B767, which seats seven abreast in the economy cabin in a 2-3-2 configuration, with 18-in. seat bottoms. Assuming our example aircraft has a 17.2-in. seat bottom, the clearances will have to be adjusted for the seat width difference between a 17.2-in. seat bottom and an 18-in. seat bottom. This is done as follows.

The equation governing the calculated clearances is as follows:

$$\text{New clearance} = \text{Baseline clearance} + (\text{New seat bottom} - 18) \times 0.75 \tag{3.3}$$

As an example for 17.2 in. seat bottoms, the new head clearance would be

$$\begin{aligned} &= 7.3 + (17.2 - 18) \times 0.75 \\ &= 6.7 \text{ in.} \end{aligned}$$

Note: The above dimensions are guidelines only. Reasonable variations in the dimensions should be explored to arrive at the optimum cross section (one that is both comfortable and has a low perimeter that allows for a low-drag fuselage).

Using the center of the outboard seat as a guide, draw in clearance circles as follows:

- Floor level—radius of ~9 in. (allows for clearance of floor air vents)
- 25.1 in. above floor level—radius of 11.5 in.
- 36.5 in. above floor level—radius of 10.2 in.
- 49.5 in. above floor level—radius of 6.9 in.

The height above the floor level is set by human factor guidelines shown in References [23, 25, 28, 29] and corresponds to the location of the feet, arms, shoulders, and head, respectively, of a seated person.

Draw in a point along the same seat centerline, about 62 in. from the floor. This will make sure there is adequate room above the passenger's head to fit in an overhead bin with a passenger service unit (lights, oxygen outlets, etc.). This is the minimum recommended clearance height. There are aircraft out there with values that are a little lower, but this number reflects good design practice. As the number of seats abreast increases, this becomes less of a constraint and vice versa.

FIGURE 3.3 Applying clearance values to the seat.

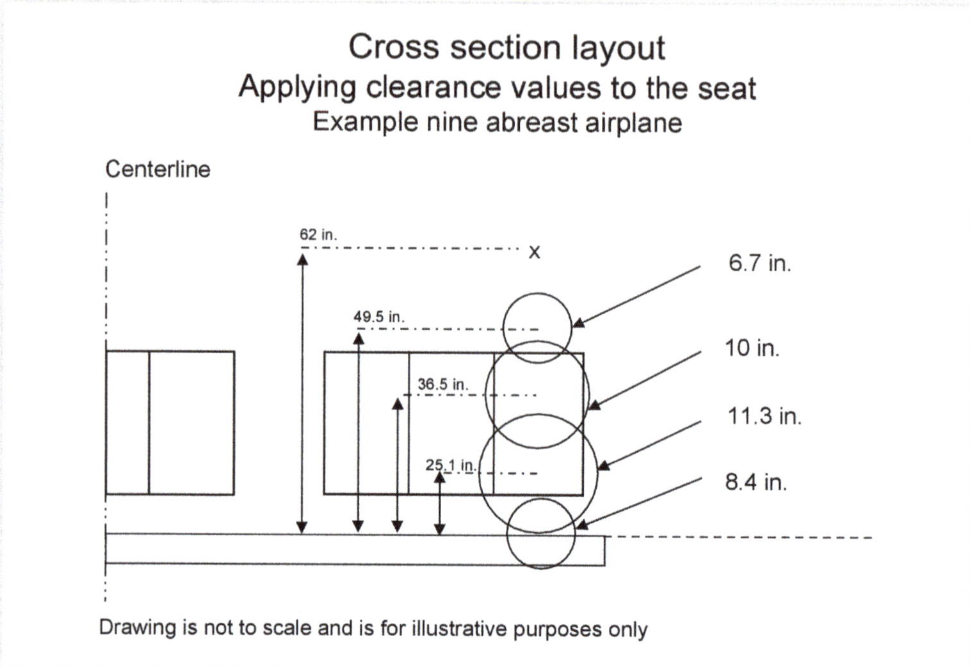

Below the floor, draw in the front view of the required container, which in our case is an LD-3 type container. Since we are focusing our drawing on the right-hand side of the airplane, we will locate the top inner corner of our container so that it is 4 in. below the floor and 1 in. away laterally from the centerline.

Draw a 2-in.-radius circle such that its center is located on the bottom right corner of the LD-3. The purpose of the circle is to establish a clearance envelope for systems and structure when drawing the cross-section lower lobe.

In our configuration, we are not considering any overhead space above the cabin for a crew rest area or a potential work or storage area. If we wanted to, we would need to make sure that there is adequate vertical height (above the main cabin floor) for the appropriate usage. A simple rule of thumb would be at least 142 in. for a crew rest area, with perhaps an additional 15 in. of height if the area was to accommodate a full-standing crew member. These measurements are from the top of the passenger floor to the inner mold line (IML) of the cross section along the centerline.

We are now ready to draw the inner mold line of our cross section. The basic idea is to wrap as tight of a curve (typically a conic) around the upper and lower lobes. Ideally, the cross section would be a circle (for structural reasons), which encompasses both the lobes, such as in the case of the 777 or the A330 or A340 line, but very often it is a double bubble, where the upper and lower lobes are attached at the passenger floor (such as in a B767). Depending on the manufacturing method that will be used, it may also be desirable to have the two lobes have a continuous curvature between them.

The upper lobe in our case is drawn by making a three-point arc, which is tangent at the centerline and touches the foot clearance and head clearance bubbles. This is the inner mold line. Different clearance bubbles may be the sizing dimensions based on the number of seats abreast and the crown height (height of the IML above the floor).

A parallel or offset curve is drawn at an offset of 1 in. to represent the cabin's inner liner that separates the passenger cabin from the frames, insulation (thermal and noise), and outer skin. This offset circle is actually the inner radial dimension of the frame.

The lower lobe is also drawn by making a three-point arc, which is tangent at the centerline and touches the foot clearance and LD-3 container clearance bubbles. Here, care must be taken to make sure that the lower lobe curve meets up with the offset outer curve and not the inner liner. This is because a similar liner does not exist for the lower lobe. Note that the measurement outlined here applies to this cross section only and will vary based on the number of seats abreast.

The outer mold line of the cross section (upper and lower lobe) is drawn at an offset, which is consistent with the frame depth—as calculated during the sizing phase.

FIGURE 3.4 Defining the liner, frame, and lower lobe details.

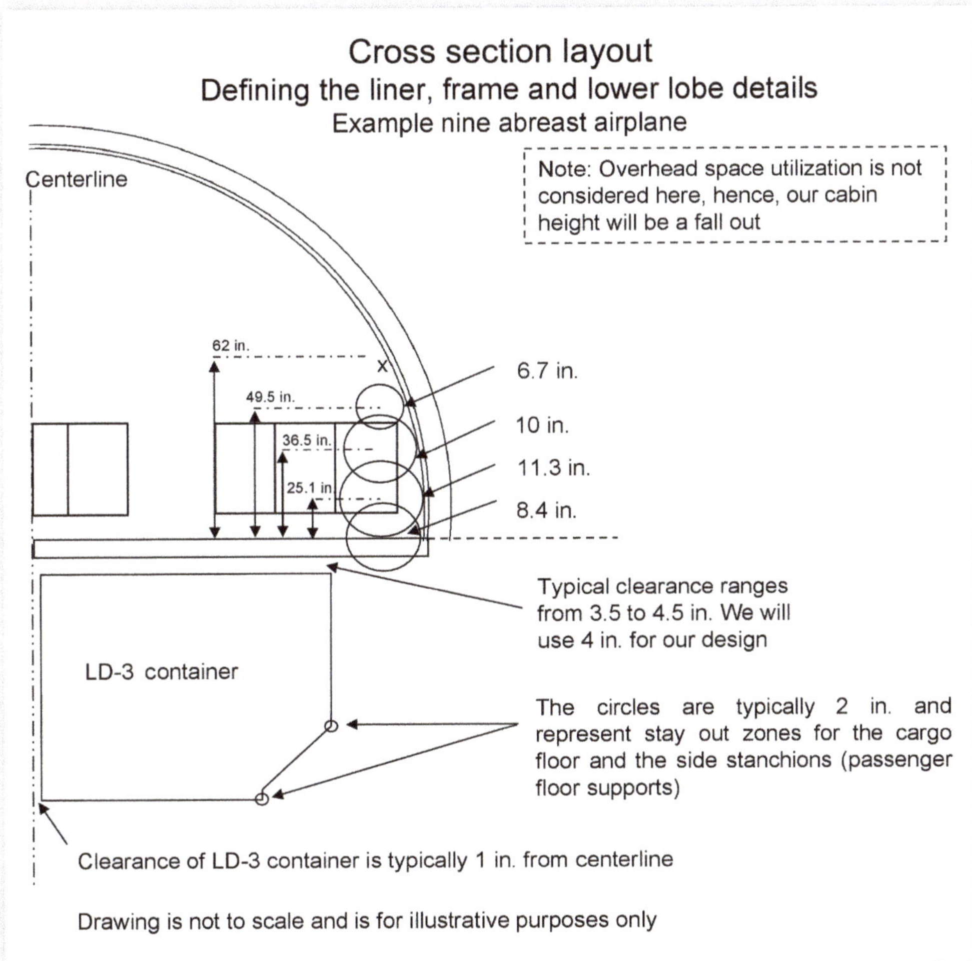

FIGURE 3.5 Completing the cross-section layout.

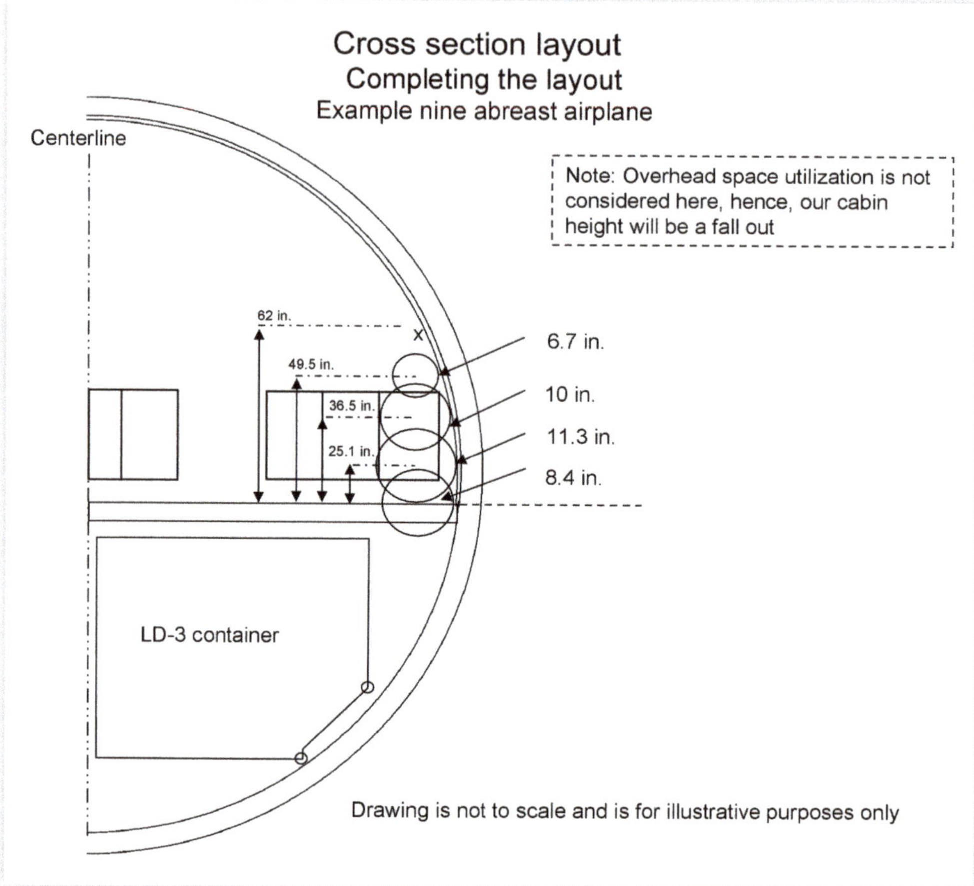

The cross section that you end up with may not strictly adhere to the clearances prescribed in the table for 17.2-in. seat bottoms. This may be the outcome of an effort to achieve as tight a cross section as possible. As long as the dimensions fall within the minimum acceptable clearances, the concept will have achieved a robust start.

There could also very well be a difference in perimeter and hence the wetted area between the calculated values during sizing and the as-drawn values. A difference in perimeter of less than 3% can be designed out in the preliminary design phase. After all, we still have to draw the wing, empennage, and wing-to-body fairing where there will be plenty of opportunities to "tighten" things up.

We now have the maximum cross-section height and width, which will help us lay out the fuselage.

Cross Sections of Existing Jet Transports

FIGURE 3.6 Single-aisle cross-sectional data.[1]

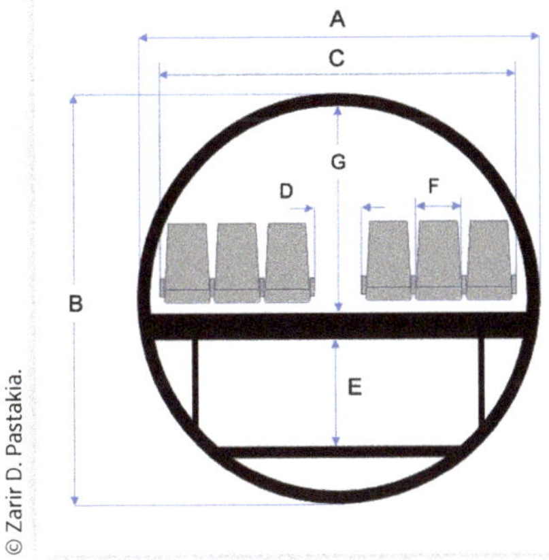

Single aisle aircraft

Model	A	B	C	D	E*	F	G	Seats abreast in economy
E175/190 E2	119	132	106	19.0	37	18.0	79	4
A220	138	146.5	129	20.0	42.5	18.5**	86 est. offset aisle	5
A320	155.5	162.97	143	19.0	45.0	18.0	94 est.	6
B737	148	158	138	20.0	45.5***	17.2	87 est.	6
MC-21	160	160	148	24.0	48 est.	18.0	92 est.	6

Note: All dimensions are in inches
Arm rest widths are 2 in. unless called out

*Top of cargo floor or roller to ceiling liner
** Dimension applies to 4 out of 5 seats. The middle seat is 19 in.
*** Average dimension between forward and aft cargo compartment

[1] Reference [72, 73, 77].

FIGURE 3.7 Twin-aisle single-deck cross-sectional data.[2]

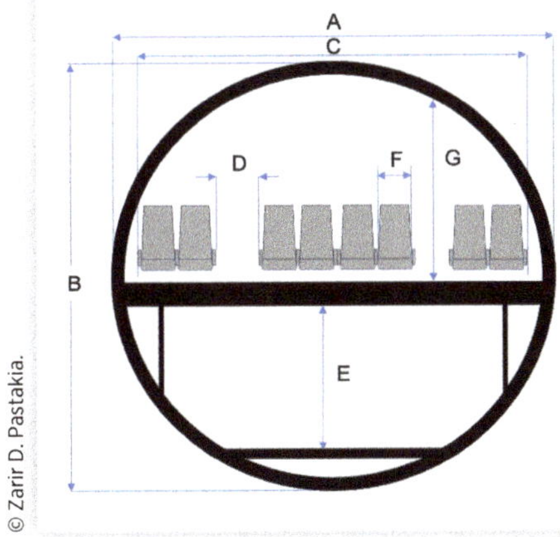

Twin aisle single deck aircraft

Model	A	B	C	D	E**	F	G	Seats abreast in economy
A330	222	222	204	19.0	66.9	18.0	Est. 123	8
A350-XWB	234.6	239.8	216.5	18.1	69.2	17.7*	Est. 125	9
B767	198	213	184	19.0	66.0	18.0	Est. 109	7
B777 classic	244	244	230	17.0	67.0	17.0	Est. 129	10
B777-9	244	244	234	18.0	67.0	17.20 17.5*	Est. 129	10
B787	226.5	234	214.8	18.0	67.0	17.20 17.5*	Est. 118	9

Note: All dimensions are in inches
Arm rest widths are 2 in. unless called out
Aisle height from floor to frame is a fall out

*Estimated using an arm rest width of 1.75 in.
**Top of cargo floor roller to ceiling liner

[2] Reference [72, 73].

FIGURE 3.8 Twin-aisle double-deck cross-sectional data.[3]

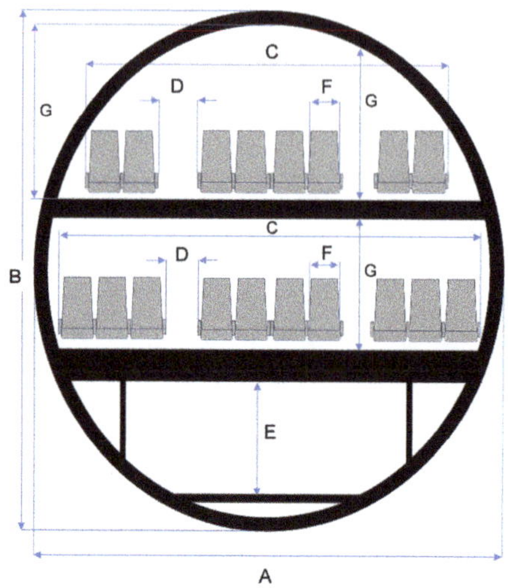

Twin aisle double deck aircraft

Upper deck dimensions

Model	A	B	C	D	E**	F	G	Seats abreast in economy
A380	281.2	331 fwd 337 aft	208	20.0	N/A	18.0	Est. 97	8
B747-8i	N/A	303	136	17.0*	N/A	17.2	Est. 85	6

Main deck dimensions

Model	A	B	C	D	E**	F	G	Seats abreast in economy
A380	281.2	331 fwd 337 aft	248	20.0	68.9	18.0	Est. 94	10
B747-8i	256	256***	239	19.5	66.0	17.2	Est. 97	10

Note: All dimensions are in inches
Arm rest widths are 2 in. unless called out

*The 747 has a single aisle on the upper deck
**Top of cargo floor roller to ceiling liner
***The aft fuselage of the 747 has a single deck which is circular

[3] Reference [72, 73].

Tamme Wichmann/Shutterstoc

4

Fuselage Layout

Developing the Fuselage Plan View

At this point, we have the cross-section drawn up, which will help us define the maximum width and height to start the fuselage layout process. Starting with the plan view or, top view is the easiest and helps to rapidly establish the fuselage length and width, utilizing our estimate from the sizing exercise and cross-section drawing, respectively.

FIGURE 4.1 Fuselage layout, top view.

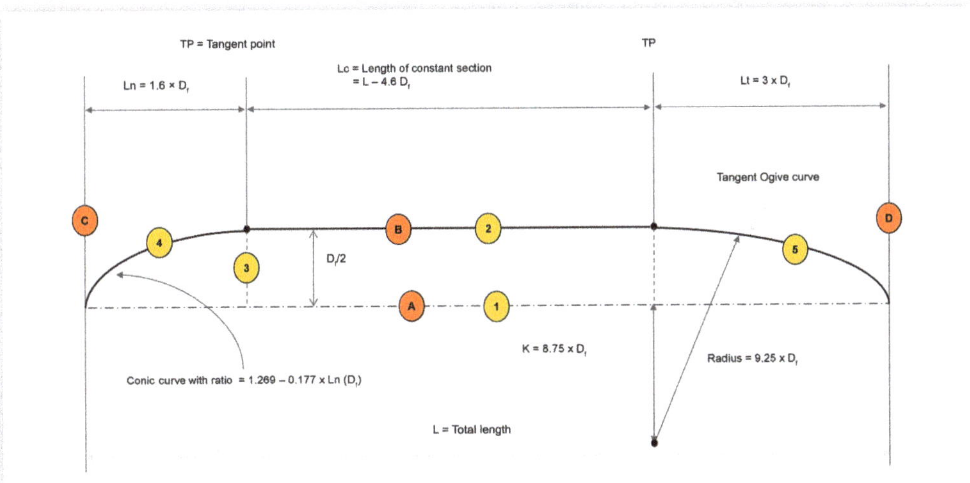

FIGURE 4.2 Top view, details of the forward and aft nonconstant fuselage sections.

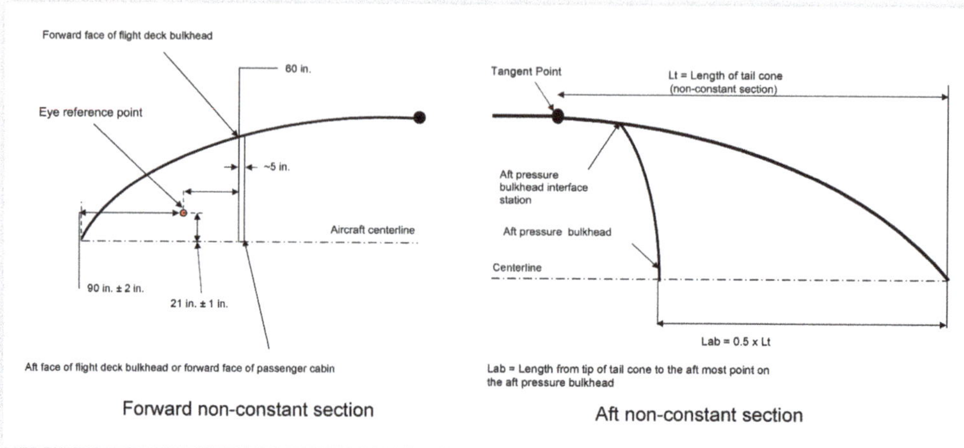

1. Draw a centerline of length "L" equal to the fuselage length from the sizing exercise. Curve A.
2. Offset a line of length "L_c" at a distance equal to half the maximum diameter "D_f" of the fuselage. Curve B.
3. Curve B should start at a distance "L_n" from the nose of the airplane.
4. Draw a conic curve that is tangent to curve B and curve C. The conic ratio is defined by the equation shown in the figure. The conic curve is intended to give a flight deck width, which is acceptable in terms of comfort, with a reasonable window size.
5. Draw the aft nonconstant section (an Ogive curve) such that it is tangent to curve B. The graphical method of drawing a tangent Ogive is shown in the figure. Note, the choice of an Ogive curve is arbitrary, and one can fit a conic or spline curve as long as the curve is tangent to the constant section. In the end, the correct fullness of the curve will be a trade between weight, drag, and the ability to get in an extra seat abreast or two in the tapered section.
   - The pilot eye reference point typically lies about 90 in. ± 2 in. aft of the nose of a single-deck airplane. The eye reference point is offset ~21 in. ± 1 in. from the centerline of the fuselage. The forward face of the flight deck partition is typically 60 in. aft of the eye reference point. Recent security requirements make it necessary to have a flight deck partition that is about 5 in. thick. This forms the aft face of the flight deck and the start of the passenger cabin.
   - The aft part of the cabin is defined by the location of the aft pressure bulkhead. The aft pressure bulkhead can be a dome or flat structure. It is essential that, for optimum fuselage structural weight, the fuselage cross section be circular where the aft pressure bulkhead interfaces with the frame station. An initial estimate for the aft pressure bulk location is 0.5 × the length of the nonconstant section.

- The aft face of the flight deck partition and the aft pressure bulkhead define the boundaries of the cabin and at the floor level. The floor area can now be checked to ensure we have sufficient cabin length to accommodate the required seat count. Since the tri-class or dual-class seat count is known, the square footage of the floor area should be within 1% of the calculated values below. Make sure the floor area calculated is done so offset from the frame depth and interior liner.

FIGURE 4.3 Required cabin floor area.

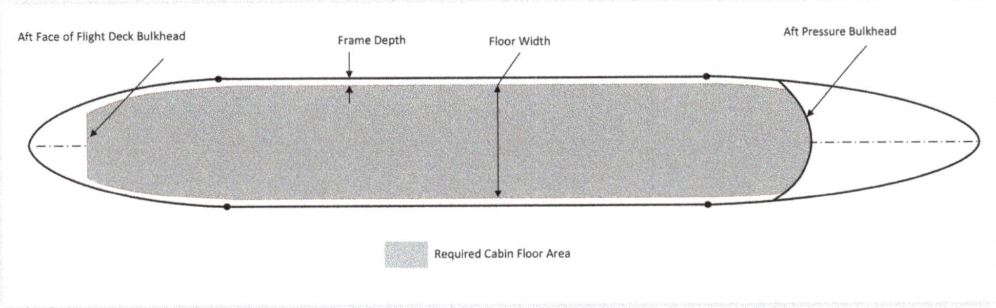

- If the calculated cabin floor area is not within the tolerance of the required floor area, then the constant section length should be adjusted (increased or decreased as necessary) as follows:

$$\text{Delta fuselage length} = \{\text{Cabin area required} - \text{Cabin area calculated}\} \div \{\text{Maximum floor width from inside of frame}\} \quad (4.1)$$

Developing the Fuselage Side View

1. Having drawn the plan view of the fuselage, we can now go on to define the side view. The constant and nonconstant lengths on the side view are the same as those on the plan view. The first step in defining the curves of the side view is to establish the maximum half-breadth waterline and the floor of the passenger cabin. The maximum half-breadth line is synonymous with the waterline of the maximum fuselage width as represented below by curve E. In wide-body transports, it is not unusual for the maximum fuselage half-breadth line to coincide with the floor waterline, but this is not always the case. The location and thickness of the floor are obtained from the cross-section drawing.
2. Define the constant section curves that are offset the appropriate distance from the floor waterline. Reference the fuselage cross-section drawing to establish the correct offsets from the floor waterline.

FIGURE 4.4 Developing the fuselage side view.

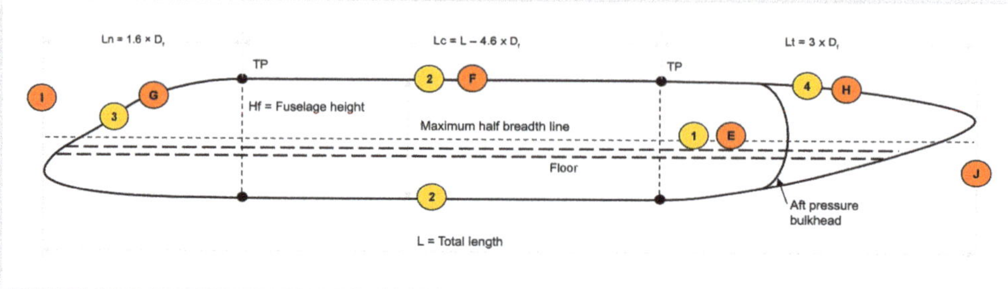

3. Draw a line that starts from the intersection of the maximum half-breadth line and the tangent point and angle down about 3 to 4°. This line will intersect the curve "I" at the given distance Ln to form the nose point of the aircraft. Locate the eye reference point 90 in. ± 2 in. aft of the nose and 45 in. ± 2 in. above the top of the floor. The outer mold line of the flight deck should be located about 23 in. ± 1 in. above and 34 in. ± 3 in. ahead of the eye reference point position. These dimensions ensure adequate room for overhead and glare shield panels, respectively. The angle of the forward windshield should typically not exceed 45° for good vision quality.

Note: In some cases, it is possible that the flight deck floor is located on a step up to 7 in. above or below the passenger cabin floor. In this case, the eye reference position should be located relative to the flight deck floor. Approximate reference point positions can be found in the airport planning documents from Boeing and Airbus aircraft. These are available publicly on the Web.

4. The next phase is to outline the approximate locations of windows and window posts to meet the minimum vision criteria and not have the pilot's vision obstructed. Remember Advisory Circular 25.773-1 specifies the minimum acceptable vision polar. In a nutshell, safety should never be compromised, and every opportunity should be taken to enhance safety where possible. The key elements of the advisory circular are listed in a future paragraph.

FIGURE 4.5 Side view, details of the forward fuselage nonconstant section.

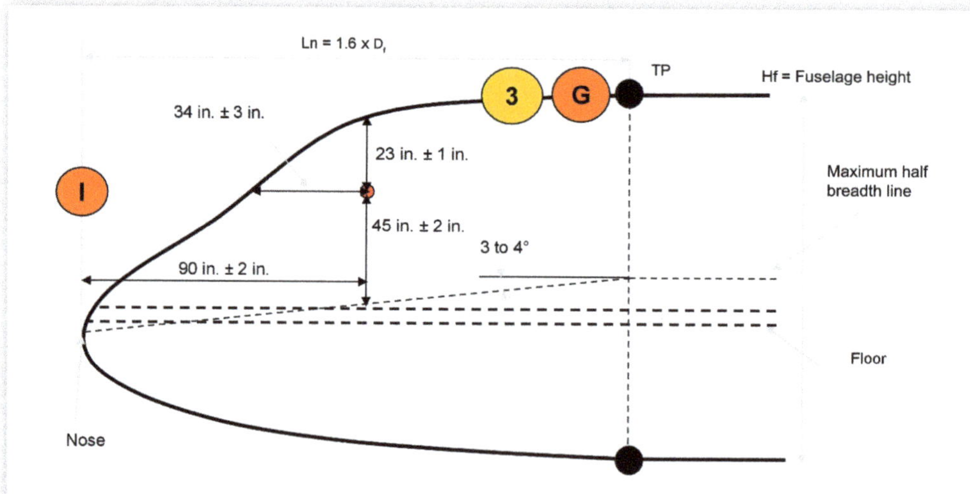

5. Draw a line that starts from the intersection of the maximum half-breadth line and the tangent point and angles up about 3 to 5°. This line will intersect the curve "D" at the given distance Lt to form the tail cone endpoint of the aircraft. The tail cone will probably end above and below this point to provide an exhaust outlet for the auxiliary power unit (APU). In the case of a blade tail, the exhaust can port through the side like on a B777 or straight out the back like on an A380. In all cases, the fuselage upsweep angle should be kept to less than 5° to reduce aft body pressure drag to acceptable levels. In unique cases, such as in cargo transport aircraft, this upsweep angle will probably be determined by the necessity for an aft loading door with some required clearances such as in an AN124, C5 Galaxy, or C17.

FIGURE 4.6 Side view, details of the aft fuselage nonconstant section.

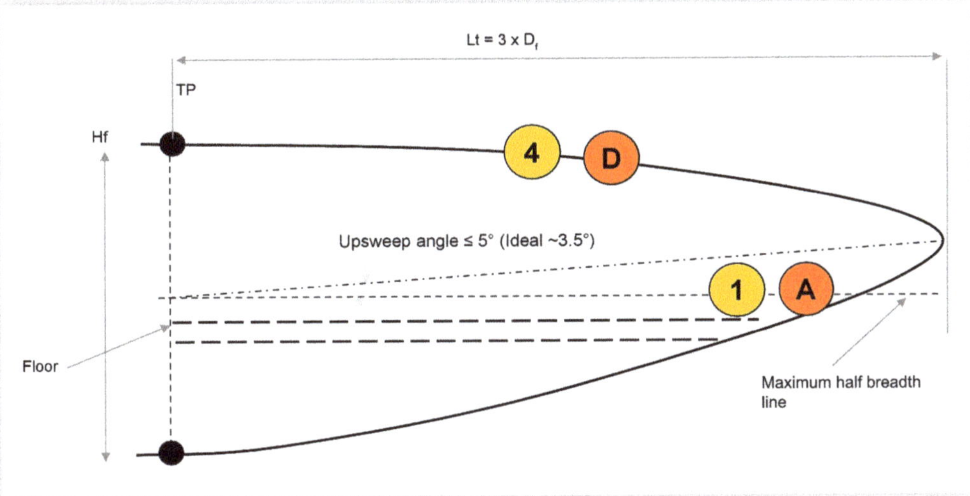

AC 25.773-1: Criteria for Pilot Compartment Visibility

- The flight deck windshield must provide sufficient external vision to permit the pilot to safely perform any maneuvers within the operating limits of the aircraft and, at the same time, afford an unobstructed view of the flight instruments and other critical components and displays from the same eye position. The following subparagraphs describe the minimum criteria for the pilot compartment view. Aircraft designers and manufacturers should make every effort to build windshields that offer the pilot more external vision.
- Design Eye Position. The design eye position is a single point selected by the applicant that meets the requirements of 53 25.773(d) and 25.777(c) for each pilot station. The figure depicting the pilot compartment view depicts a design eye position for optimum collision avoidance potential for the left pilot seat. For the right pilot seat, all left or right dimensions are reversed.

- Clear Areas of Vision. The clear areas of vision should be determined by the measurement of angles from the design eye position, utilizing binocular vision. Binocular vision is the total area that can be seen by both eyes. It is not limited to the binocular field but includes, in addition, monocular regions visible to the right eye, but not to the left, and vice versa.
- The horizontal and vertical vision angles should be measured from the following:
   i. A vertical datum plane running fore and aft through the design eye point and the central axis.
   ii. A horizontal datum plane perpendicular to the vertical plane that also passes through the design eye point and central axis. The vertical and horizontal datum planes are fixed relative to the airplane and should be parallel to those corresponding to zero pitch and yaw angles for the airplane.
- With the design eye position located, the vision through the transparent areas should provide the following pilot compartment view:
   i. Forward and up 35° from the horizontal datum plane at 40° left of the vertical datum plane, diminishing linearly to 15° up at 20° right.
   ii. Forward and down 17° from the horizontal datum plane between 30° left and 10° right of the vertical datum plane, diminishing linearly to 10° down at 20° right.
   iii. Forward and up 35° from the horizontal datum plane between 40° left and 80° left of the vertical datum plane, diminishing linearly to 15° up at 120° left.
   iv. Forward and down 17° from the horizontal datum plane at 30° left of the vertical datum plane, diminishing linearly to 27° down at 70° left.
   v. Forward and down 27° from the horizontal datum plane between 70° left and 95° left of the vertical datum plane, diminishing linearly to 15° down at 120° left.

FIGURE 4.7 Pilot compartment vision polar.[1]

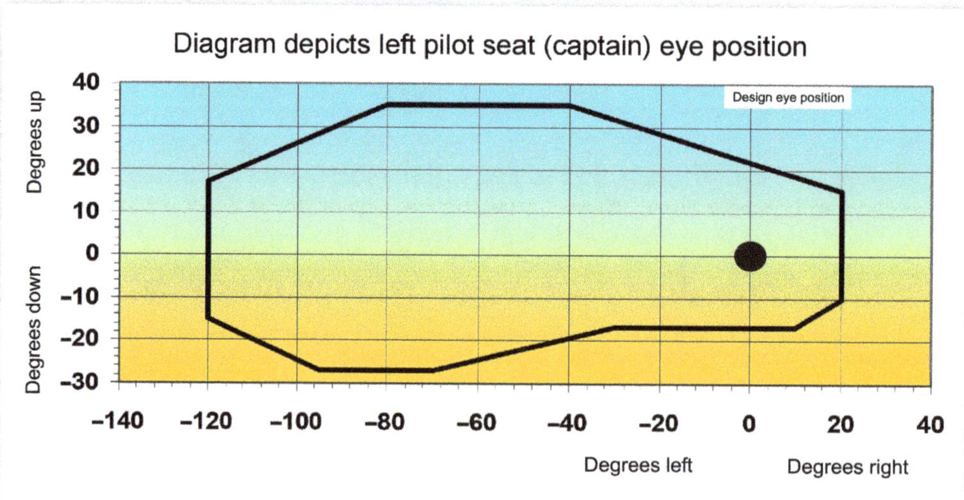

Federal Aviation Administration.

[1] https://www.faa.gov/documentlibrary/media/advisory_circular/ac_25_773-1.pdf

FIGURE 4.8 Vision polar side and top view.

Vision polar from the flight deck (side and top view)

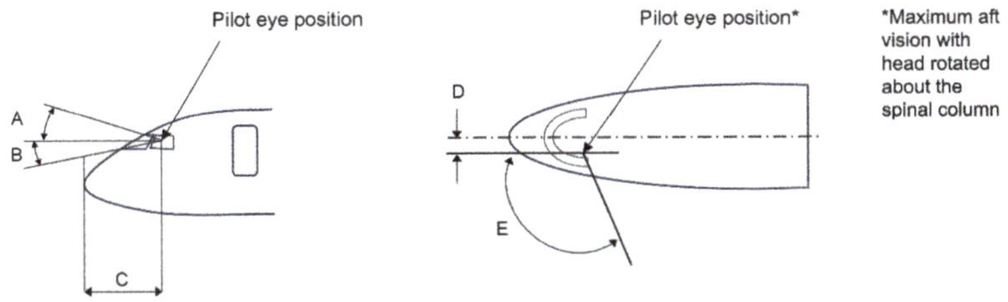

Aircraft	A	B	C	D	E
Embraer 190	27.8°	15°	115 in.	21 in.	120.6°
A320neo	30°	20°	94.9 in.	21 in.	111°
Boeing 787	25.4°	21°	21 in.	21 in.	121.5° Eye position aft by 3.3 in.
Boeing 747	22°	18.8°	218 in.	21 in.	123° estimated

Note: The vision out the side window can be improved by up to 22° (geometry dependent) when the pilot eye position moves 5 in. further outboard. This is sometimes called the alert position.

FIGURE 4.9 Vision polar front view.

Vision polar front view

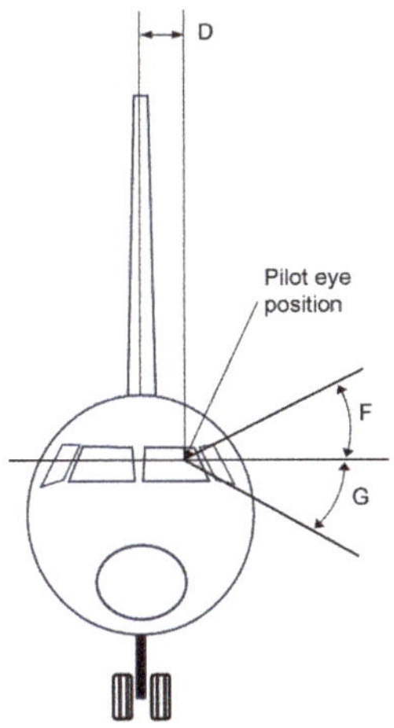

Visual angle in plane perpendicular to longitudinal axis through pilot's eye position

Aircraft	D	F	E
Embraer 190	21 in.	27.5°	28.1°
A320neo	21 in.	33°	32°
Boeing 787	21 in.	31° Pilot eye position moved aft 3.3 in.	27° Pilot eye position moved aft 3.3 in.
Boeing 747	21 in.	37°	34°

Note: The vision out the side window can be improved by up to 22° (geometry dependent) when the pilot eye position moves 5 in. further outboard. This is sometimes called the alert position.

Landing Vision

The view angle forward and down should be sufficient to allow the pilot to see a length of approach and/or touch-down zone lights that would be covered in 3 sec at landing approach speed when the aircraft is as follows:

1. On a 2.5°-glide slope.
2. At a decision height that places the lowest part of the aircraft at 30.5 m (100 ft) above the touch-down zone extended horizontally.
3. Yawing to the left to compensate for 10 knots of crosswind.
4. Loaded to the most critical weight and center of gravity.
5. Making the approach with 366 m (1200 ft) of runway visual range (RVR).

Obstructions to Vision

There should be no obstructions to vision between 20° right and 20° left in the vision polar depicted in the figure. Obstructions outside this 40°-area should be kept to a minimum, ideally not more than three (i.e., center post, forward post, and side post). Using binocular vision, it should be possible for a pilot to have a vision of any given bearing that is blocked to the other pilot from 80° right to 80° left of the design eye position. In addition, it is desirable that obstructions be eliminated by using binocular vision with the average human intraocular dimensions of 63.6 mm (2.5 in.). The use of sun visors that reduce light transmissivity is acceptable; however, totally opaque visors that impinge upon the field of view of Figure 4.7 should not be used.

Chesky/Shutterstock

5

Cabin Layout and Door Positioning

The cabin and the cargo compartment are the revenue-generating elements of the airplane. It is the payload-carrying element of the airplane that defines how successful the design will be in the marketplace. For military airplanes, this is the armament such as rockets, missiles, bombs, and surveillance equipment. In commercial airplanes, it is typically passengers, bags, and revenue-generating cargo. In short, the load which generates money to travel from point A to point B.

A good design process ensures that the "as-drawn" airplane has synergy with the initial sizing assumptions. In the initial sizing phase, the floor area per seat metric enables a quick calculation of fuselage length. It establishes a target value that reflects the state of the art in cabin design, such as space-saving lavatories, seat designs, and galleys. Defining a cabin arrangement to achieve this metric of floor area per seat is complex and governed by regulatory requirements that are defined by different certification authorities such as the Federal Aviation Administration (FAA) and the European Union Aviation Safety Agency (EASA). Different cabin product offerings between "no frills" operators such as Ryan Air versus a full-service mainline carrier such as British Airways further add to the challenge of cabin definition. Several cabin layout tools are available today that have most regulations built into the rule set and can be used to establish a reasonable start point for an airline cabin layout.

However, at this initial stage in the design, we can start with the type of exits and how they can be distributed. The guiding Federal Aviation Regulations (FARs) are FAR 25.807 and Advisory Circular (AC) AC 25-807-1. This approach combined with a good comparison of reference commercial jetliners in operation today helps lay the foundation for a reasonable start point.

Simplistically, you want the minimum number of exits necessary to satisfy the marketing requirement for seats. Narrow-body or single-aisle aircraft can have significant variability in seating layouts ranging from single-class to dual-class and, in some cases, even three-class layouts that include premium economy. However, the single-class layout typically drives the maximum seating capacity and sets the number and type of exits that have to be planned for in a design. In a wide-body or twin-aisle aircraft, it is the dual-class layout that drives the number and type of exits; however, in unique cases such as the Japanese domestic market, short-range missions are flown in single-class high-density layouts and can influence exit types. A good example of this is the use of B777 and B747 aircrafts used by ANA and JAL on short 1- and 2-h domestic mission stage lengths.

Understanding the Rules and the Terminology

Passenger Cabin Length

- Typically, airline manufacturers define the cabin length as the distance between the flight deck bulkhead and the aft pressure bulkhead that constitutes the usable floor space on a transport. This is also the definition used to calculate the floor area per seat number we use in the initial sizing phase.
- However, for the purposes of laying out an airplane and meeting the rules prescribed in the FARs, the exact definition is described as the distance from the centerline of the forwardmost exit to the centerline of the aftmost exit in the airplane. If there are four or more rows of passenger seats located beyond the most forward or most aft exits, then the cabin starts or ends at the centerline of the front stud of the most forward of most aft passenger seat.

Passenger Zones or Zones

- A section of the passenger cabin which is bound longitudinally (along the length of the cabin) by a pair of exits on both ends.
- In instances, where there are seats installed beyond the most forward or most aft pair of exits, a zone can also be defined as a section of the passenger cabin bounded by the start or end of the passenger cabin and the nearest pair of exits.

Exit Rating

The increase in passenger seating configuration allowed by the installation of a pair of that "type" of exit. "Types" of exits and exit pairs are described in Table 5.1.

TABLE 5.1 Types of exits.

Exit	Seat rating per exit pair	Opening (in.) Width × Height × Corner radius	Floor Level
A	110	42w × 72h × 7r	Yes
B	75	32w × 72h × 6r	Yes
C	55	30w × 48h × 10r	Yes
I	45	24w × 48h × 8r	Yes
II	40	20w × 44h × 7r	Yes, unless overwing. Then, step up 10 in. inside the cabin and step down 17 in. onto the wing
III	35	20w × 36h × 7r	Yes, unless overwing. Then, step up 20 in. inside the cabin and step down 27 in. onto the wing
V	9	19w × 26h × 6.3r	Yes, unless overwing. Then, step up 29 in. inside the cabin and step down 36 in. onto the wing

Exit Unit

A dimensionless number that is related to the exit rating of a pair of exits of the same type and uses the rating of the type III exit (35) as the baseline (Table 5.2). Hence, the exit unit value of a pair of exits is determined by dividing the rating of the exit by 35 and rounding the value down to the next quarter of a unit. For example, a type A exit rating has a unit number of 3.0 (110 ÷ 35 = 3.14 rounding down to the nearest 0.25 = 3.0).

TABLE 5.2 Exit type and exit unit.

Exit type	Exit unit
A	3.0
B	2.0
C	1.5
I	1.25
II	1.0
III	1.0
Two type III exits within three seat rows of each other	2.0

Fuselage Length Factor

A length is determined by dividing the length of the passenger cabin by the sum of the exit units in each zone of the airplane.

Exit Offset

Exit offset is the distance between the proposed exit centerline and the calculated exit centerline. The offset value allowed is ±15% of the total length of the cabin. As an example, if the cabin is 2000 in. in length, then exits can be offset from their nominal values by 300 in.

The nominal exit location and the allowed ±15% deviation from that location are not applicable when they position the exit in an area that could be hazardous to passengers during an emergency evacuation, such as in the vicinity of propellers, engine air intakes, or pylons, landing gear, and the wing leading and trailing edges.

Also, no two exits can be more than 60 ft or 720 in. apart from each other when measured door edge to door edge.

Individual Zone Passenger Capacity

The maximum number of passengers that may be seated in an individual zone is the sum of the ratings of the exits that bound the zone.

Sequential Zone Passenger Capacity

The maximum number of passengers that may be seated in two or more adjacent zones is the sum of the ratings of the exits that are either bound or included within the adjacent zones.

FIGURE 5.1 Example calculation for exit distribution.[1]

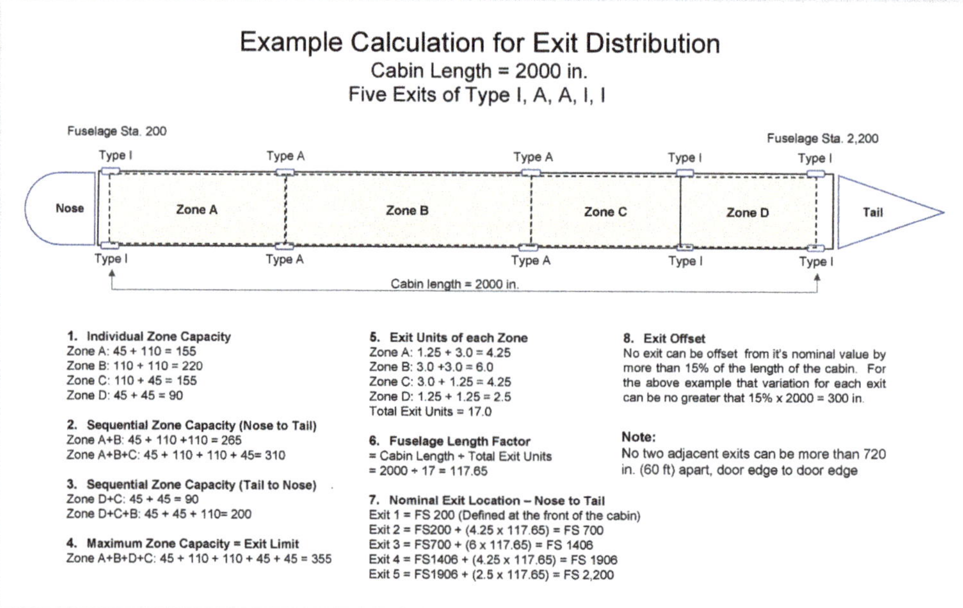

[1] https://www.faa.gov/documentLibrary/media/Advisory_Circular/AC_25_807-1.pdf.

FIGURE 5.2 Single-deck airliner exit type and distribution.[2]

Single Deck Airliner - Exit Type and Distribution

	Typical Dual Class Seating Capacity *Book Method Est.*	**Exit Rating & Distribution (Typical Seating Capacity)** *Exits are re-rated or de-activated to take advantage of cabin space and single lane slides where applicable*	**Exit Type & Distribution (Max. Passenger Seating Capacity)** *Exits are sized and / or optionally activated to maximize seating capacity for higher density operations or planned future growth*
A220	129	C – III – C	C – III – C (145)
A320	147	C – III – III – C	C* – III – III – C* (195)
A321	193	C – C – C - C	C* - (III-III)+ - C - C* (244)
A350-900	319	C – A – C – A	A – A – A – A (440)
A350-1000	372	C – A – A – A	A – A – A – A (440)
B737 MAX 8	159	I – III – III – I	I – III – III – I (189)
B737 MAX 10	186	C – III – III – III – C	C – III – III – C – C (230)
B767-300	214	A – III – III – A	A – A – I – A (351)
B787-9	293	C – A – C – A	A – A – A – A (420), (exits accommodate up to 440 in a 787-10)
B777-300ER	386	C – A – A – A – A	A – A – A – A – A (550)
B777-9X	414	A – A – A – A	A – A – A – B – A (515)
E2-190	95	I – III – I	I – III – I (114)

* Over sized / Over performing Type C exits that get credit for 10 more seats per Type C exit (going from 55 seats to 65 seats)
\+ Over sized / Over performing Type III over wing *exit pairs* that get credit for 5 additional seats per exit pair (going from 65 to 70 seats)

[2] https://www.faa.gov/documentLibrary/media/Advisory_Circular/AC_25_807-1.pdf.

FIGURE 5.3 Double-deck airliner exit type and distribution.[3]

Double Deck Airliner - Exit Type and Distribution

	Typical Tri Class Seating Capacity *Book Method*	**Exit Rating & Distribution (Typical Seating Capacity)** *Exits are re-rated or de-activated to take advantage of cabin space and single lane slides where applicable*	**Exit Type & Distribution (Max. Passenger Seating Capacity)** *Exits are sized and / or optionally activated to maximize seating capacity for higher density operations or planned future growth*
B747-8	421	Upper Deck – I Main Deck – C – A – A – A – C	Upper Deck A (110) Main Deck C – A – A – A – A (495) Total Passenger Seating (605)
A380-800	530	Upper Deck – A – A – A Main Deck – C – A – A – A – C	Upper Deck A – A – A (330) Main Deck A – A – A – A – A (538) Total Passenger Seating 868

Note: Given the height of the upper deck, it is not clear if a single lane slide would be an acceptable solution for human factors reasons, hence it was estimated that on the 747-8 and the A380 the upper deck doors, while re-rated would likely retain a dual lane slide for passenger evacuation. The relevance to the aircraft design is weight. While re-rating can be a benefit to cabin space and passenger count, it can also help to reduce weight, by going from a dual lane slide to a single lane slide; such as, in the case of going from a Type A rating to a Type C rating.

[3] https://www.faa.gov/documentLibrary/media/Advisory_Circular/AC_25_807-1.pdf.

Terminal Servicing and Other Considerations

The rules defined in AC 25-807 help determine how many doors and their locations need to be made available to passengers in the various zones, such that they can successfully leave the cabin in an emergency in the allotted 90 seconds. Door locations drive slide locations, which also must be considered in the context of safety in relation to objects such as rotating engines and other hazards.

FIGURE 5.4 Escape slides shown for the Airbus A380.

Anilyanik/istockphoto.com.

Besides the emergency evacuation requirements, there are also practical considerations for daily operations that require adequate room for jetways and other ground support equipment that must service the aircraft. Other doors such as those that provide access to the lower deck for cargo also must be placed such that they do not interfere with the main deck doors and aircraft structure (wing trailing edges, engines, etc.).

FIGURE 5.5 Boeing 787 Airplane Servicing Arrangement.

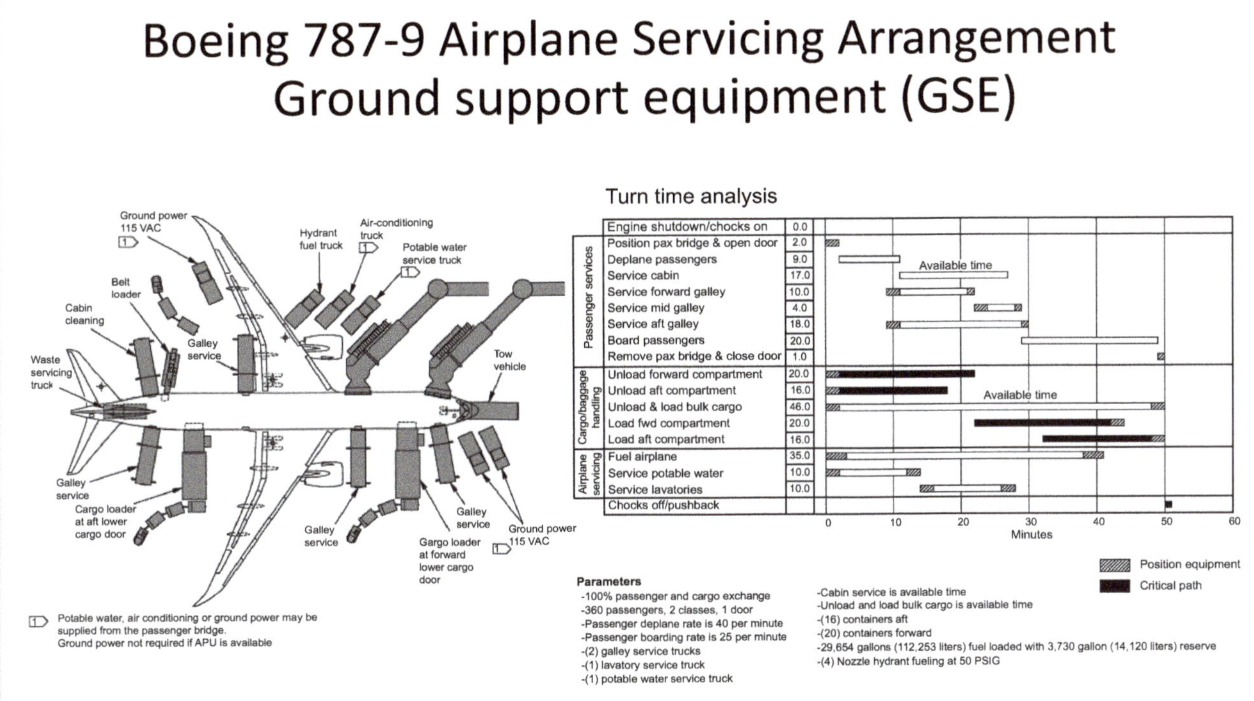

Courtesy of Boeing.

FIGURE 5.6 Door-to-door separation and door-to-engine nacelle clearances.

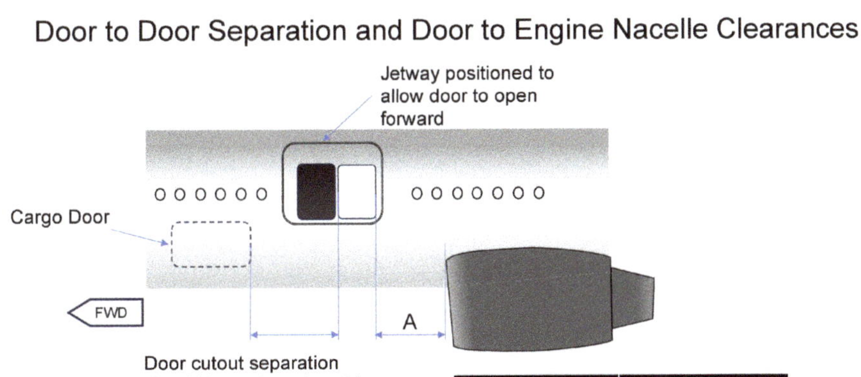

Aircraft	Dimension A (Inches)
A318	140.3
A320 neo	220.6
B737 MAX 7	209.0
B767-400ER	79.8
B777-9	77.0
A350-900	97.6
A380 (Upper Door)	29.7

© Zarir D. Pastakia.

FIGURE 5.7 An ANA Boeing 777-300ER being serviced.

Along the fuselage, the main deck doors and lower deck doors must be separated so that there is enough room around the doors to add additional material to maintain the door shape and the surrounding fuselage's structural strength. Not enough room means excessive material and that means excessive weight. The layout of the aircraft is the first step in ensuring that all these elements are adequately planned out for an efficient airframe.

Chatchai Kritsetsakul/Shuttersto

6

Wing and Landing Gear Layout

Common Wing Terminology

Aspect Ratio

Most transonic transport wing aspect ratios can range from 6.0 to 12.0. A high aspect ratio wing makes for a more aerodynamically effective, but structurally heavier wing. Aerodynamic efficiency is realized via an increase in the span of the wing for a given wing area. The increased wingspan helps with induced drag reduction in all phases of the flight. A high aspect ratio wing also increases the wing lift curve slope. Hence, for a given angle of attack the wing will generate more lift. This helps with reducing cruise drag and the thrust required for take-off and landing, both of which help improve environmental noise. This decrease in area is typically balanced against increased structural weight for the more slender surface.

Taper Ratio

Wing taper ratios for transonic jets can range from 0.2 to 0.5. A high taper ratio value can make for a more efficient but potentially heavier wing. A high taper ratio also helps to improve the induced drag of the wing by allowing more of the wing area to be biased further out. The additional area can allow the wing to have a more elliptic span load, which is good for induced drag reduction. This decrease in induced drag is typically balanced against increased structural weight due to a more slender box structure at the root. In the context of the wing layout, the taper ratio is often used to help integrate the different components such as flaps, slats, ailerons, and spoilers and their respective actuation systems. A low taper ratio, however, helps bias more area toward the inboard wing, and this can be used to better integrate the landing gear, especially on a swept wing.

Thickness-to-Chord Ratio

Thickness-to-chord ratios for transonic wings typically range from 8 to 12%. A low thickness-to-chord ratio is usually good for aerodynamics but bad for internal box volume and weight. Volume can include fuel volume and/or volume to allow for system runs.

Dihedral

Conventional jet transports usually carry a fair amount of dihedral on the wing. This is usually done to make sure that with low-wing-mounted engines there is enough clearance for the engines such that the landing gear is not unusually long. Dihedral also contributes to roll stability and has the effect of allowing fuel to flow from the tips to the center of the wing where it is pumped back to the engines. The typical dihedral angle for a low-mounted wing (B777, A330, and MD-11) is between 4 and 7°, while that of a rear-engine configuration (CRJ-700 and Fokker 70) will be 1°. The 7° of dihedral number is usually associated with four-engine transports with engines on the wing and is primarily done to make sure there is enough roll clearance during a landing maneuver. High-wing airplanes are inherently more stable and may have −1 to −3° of dihedral also known as anhedral.

Mean Aerodynamic Chord (MAC)

The center of gravity (CG) position on an aircraft is usually defined in terms of the mean aerodynamic chord of the wing. The mean aerodynamic chord (MAC) of a wing is the width of an equivalent rectangular wing in each set of aerodynamic conditions. The CG range (forward to aft) in terms of the MAC is set by the length of the fuselage (a longer fuselage means a longer cabin) to account for variations in passenger and cargo distribution.

FIGURE 6.1 Wing characteristics of some jet transports.

Wing Characteristics of Some Jet Transports
Reference 19 & 28

Aircraft Type	Wing Area (Projected) S_{ref} ft^2	Wing Span (Projected) ft	Inboard Flap Type	Outboard Flap Type	Total Flap Area ft^2 (Flaperon Included if Present)	Total Slat / Krueger Flap Area ft^2 (Total Leading Edge Device Area)	Total Spoiler Area ft^2	Total Aileron Area ft^2 (Flaperon Included if Present)
A320-200	1320	111.9	Single Slotted Fowler	Single Slotted Fowler	227.1	136.1	--	--
Fokker 70	1006	92	Double Slotted Fowler	Double Slotted Fowler	183.0	No LE Device	57	38
A330-200	3892	197	Single Slotted Fowler	Single Slotted Fowler	--	--	--	--
B777-200ER	4605	199.9	Double Slotted Fowler	Single Slotted Fowler	722.6	396.5	247.7	148.5
MD-11	3958	169.5	Dropped Hinge Vane Main Fowler Flap	Dropped Hinge Vane Main Fowler Flap	668.4	471.9	--	--

ESDU Wing Area

ESDU stands for Engineering Science Data Units and is an engineering advisory organization. ESDU provides validated engineering analysis tools to the aeronautics community among others. ESDU provides a definition of wing area. The ESDU wing as defined has the same exposed area (projected planform area) as the real wing. The wing area obtained from the sizing is the ESDU wing area.

Define the ESDU Planform

Known Quantities

- Wing area, S_{ref}
- Wing semi-span, $b/2 = y_0$
- Aspect ratio = Basic wing aspect ratio from sizing
- Initial guess at λ from sizing
- Y dimension of the side of the body, $Y_{sob} = 0.5 \times$ Max fuselage diameter
- A dihedral angle of 6° is a good guess
- Quarter-chord sweep – θ c/4 from sizing

Steps

1. Convert the wing area, Y_{sob}, and span, which are projected dimensions into the reference plane of the wing. This is done by dividing these values by the cosine of the dihedral angle.
2. Calculate the root chord.

$$C_r = [4 \times \text{Wing semi-span}] \div [AR \times (1+\lambda)] \tag{6.1}$$

3. Calculate the tip chord.

$$C_t = C_r \times 1 \tag{6.2}$$

$$\text{ESDU area (in wing reference plane)} = (C_r + C_t) \times y_0 \tag{6.3}$$

At this point in time, we are only interested in half the wing; hence,

$$\text{ESDU area (half wing)} = 0.5 \times (C_r + C_t) \times Y_0$$

$$\text{ESDU exposed half wing area} = \text{ESDU area (half wing)} - y_{SOB} \times [C_r - \tan(\theta_{LE}) + \tan(\theta_{TE})] \tag{6.4}$$

FIGURE 6.2 ESDU wing definition.

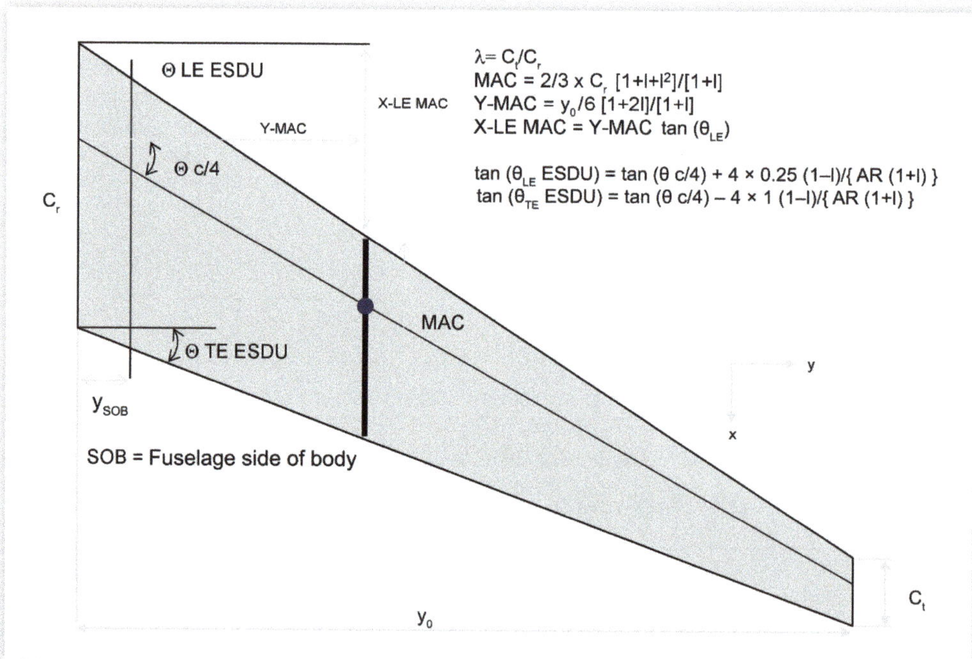

Parameterization of the Real Wing Planform

Note: All calculations are based on half the wing area.

By definition,

$$\text{Exposed wing area of the real wing} = \text{Exposed ESDU wing area}$$

$$\begin{aligned} & x_1 \times (0.5 \times y_0 - y_{SOB}) + 0.5 \times C_t \times y_0 + 0.5(x_2 \times y_1 + x_3 \times y_2 + x_4 \times y_3) - y_{SOB} \times (x_2 + x_3 + x_4) \\ & + 0.5 \times y_{SOB} \times 2 \times [\tan(\theta_3) - \tan(\theta_0)] + 2 \times x_0 \times y_{SOB} = \text{ESDU area (half wing)} \\ & - y_{SOB} \times [C_r - \tan(\theta_{LE\,ESDU}) + \tan(\theta_{TE\,ESDU})] \end{aligned} \tag{6.5}$$

Process Flow (This Can Be Solved Easily in a Program or an Excel Solver)

4. Guess at the taper ratio (use that calculated in the sizing program) if not known start with the value 0.25.
5. Guess at x_0, x_2, x_3, x_4, y_1, y_2, y_3, and y_g.
6. Solve for the sweep of the trapezoidal wing (defined by the centerline chord of x_1 and tip chord of C_t):

$$\tan(\theta_{LE}) = \tan(\theta c/4) + (1-\lambda)/\{AR \times (1+\lambda)\} - x_0 \times y_g / y_{02} \tag{6.6}$$

$$\tan(\theta_{TE}) = \tan(\theta c/4) - 5 \times (1-\lambda)/\{AR \times (1+\lambda)\} + x_0 \times y_g / y_{02} \tag{6.7}$$

7. Solve for tan (θ_0) and tan (θ_3):

$$\tan(\theta_0) = [\tan(\theta_{LE}) + x_0 / y_g] \tag{6.8}$$

$$\tan(\theta_3) = [\tan(\theta_{TE}) - (x_2 / y_1 + x_3 / y_2 + x_4 / y_3)] \tag{6.9}$$

Solve for trapezoidal wing centerline chord x_1 from Equation (6.5).

8. Locate the ESDU quarter-chord MAC point along the span at a distance of y-mac as shown in the ESDU planform drawing. At this y-location, position the MAC so that its quarter-chord point aligns with the quarter-chord point of the local wing chord.

FIGURE 6.3 Actual wing definition.

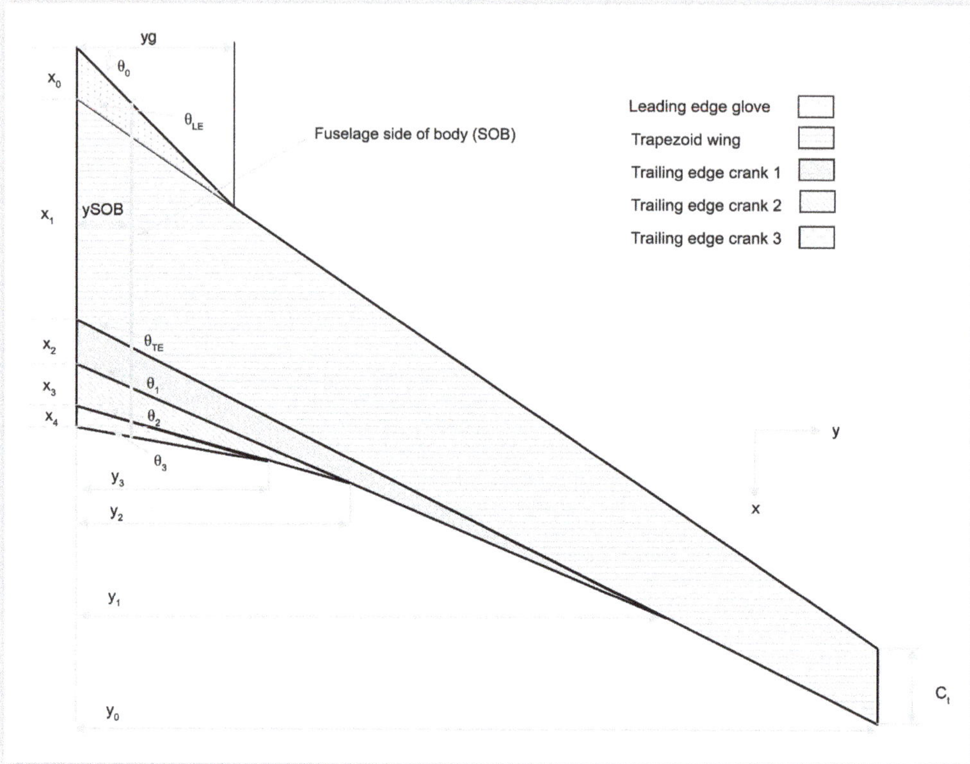

Wing Integration

1. The next step is to define integration guidelines for flaps, spoilers, leading-edge devices, engine centerlines, and the main landing gear. The aerodynamic span load is approximated to check whether the planform will give us a reasonable section cl distribution.
2. Utilize the attached charts on flap chord ratio to start laying in the rear spar centerline on the outboard wing. These flap chord ratios should be used as a guide to start the process. A given wing design can always trade flap area with wing area for a given mission specification.
3. Typical aileron chord ratios average about 21% ± 1% of the local chord. Start the aileron about 10″ inboard from the tip. Draw in a 21% chord aileron, from the TE of the planform. The inboard edge of the aileron typically ends somewhere between 68 and 78% of the span depending upon the type of aileron (use the planform drawings of existing transport wings as a guide).
4. The rear spar (RS) is offset from the LE of the aileron to allow for the aileron actuators. See the figure to guesstimate the amount of offset.
5. As we move inboard from the tip, the next constraint that sets the RS position is the flap chord ratio at the outboard end of the outboard flap. A reasonable ratio is 23% ± 2% of the local chord. The RS offset from flap LE is found in the planform guideline figure.
6. Moving further inboard, the next guiding parameter is the flap chord ratio of the inboard end of the outboard flap. This ratio is typically on the order of 27% ± 3% of the local chord. RS offset to the flap LE is found in the planform guideline figure.

7. Place the rear spar at the side of the body at the same chord-wise x-location as the quarter-chord MAC point. Placing the RS ahead of the quarter-chord is structurally inefficient for rear spar loading. The rear spar should be placed no more than 4% ahead of the quarter-chord MAC.
8. The front spar (FS) is laid out similarly starting at the tip. It will be parallel to the wing LE or have a different sweep depending upon whether the leading-edge devices (Slats and Kruger flaps) are constant chord devices or tapered, respectively.
9. Utilize the existing wing planform drawings to guesstimate the slat chord ratio at the tip. This will help define an offset for the front spar behind the leading edge of the wing.
10. The box chord at the side of body (SOB) is typically 43% ± 3% of the side of the body chord. This is a ballpark estimate. The exact number will be driven by the fuselage frame spacing, which can vary between 20 in. for aluminum airplanes and 25 in. for composite airplanes. The box chord at SOB is typically determined by lining up the RS to a frame with the FS location being a fall-out at another fuselage frame. The 43% ± 3% ratio above is intended to give the designer a reasonability check. It can be a little higher or lower depending on the frame spacing.
11. Derive approximate main landing gear (MLG) dimensions. The MLG to a large extent drives the inboard wing integration, which in turn can have implications for the outboard wing.
12. The "Y" location of the MLG on the wing can be guesstimated from the MLG sizing criteria.
13. Use the length of the MLG to set the overall trunnion length (fittings + trunnion). Locate the point on the trunnion where the "leg" (vertical part of MLG) attaches. This is ~70% of the trunnion length aft of the rear spar.
14. If the gear is raked (tilted forward or aft in the side view), then the ground, contact point of the MLG is simply [A × Sine (rake angle)], where "A" is defined in the MLG sizing criteria.
15. Check to see whether the gear ground contact point is adjacent to the 58% ± 3% ESDU MAC point.
16. At this point, a choice of a gear attach concept should be made. If a gear beam is desired, draw in a gear beam from the side of body through the tip of the trunnion as shown in Figure 6.9. The gap between the RS and the gear beam at the SOB is determined from the guidelines in the planform figure. If a cantilever gear attach concept is desired, then simply draw a line parallel to the trailing edge of the wing such that it is offset ~10 in. aft of the aft tip of the trunnion. This line will represent the spoiler beam of the inboard wing. In the case of the gear beam design, the spoiler beam will be drawn such that it is parallel to the wing TE so that it is offset about 3 to 5 in. aft of the gear beam at the side of body.
17. The leading edge (LE) of the inboard flap (at SOB) is typically offset about 3 in. aft of the spoiler beam.
18. At this point, check the flap chord ratios and see how they compare to the flap chord ratios of the existing wings, which are similar in sweep and aspect ratio to the wing you are designing. If the flap chord ratios look like they are in the ballpark and the gear ground contact point is in the correct ballpark with respect to the mean aerodynamic chord, then the planform is fairly well balanced. If not, the planform will have to be adjusted (corner points, rear spar location at SOB, and/or flap chord ratios) to get to a reasonable wing.

FIGURE 6.4 Wing planform guidelines.

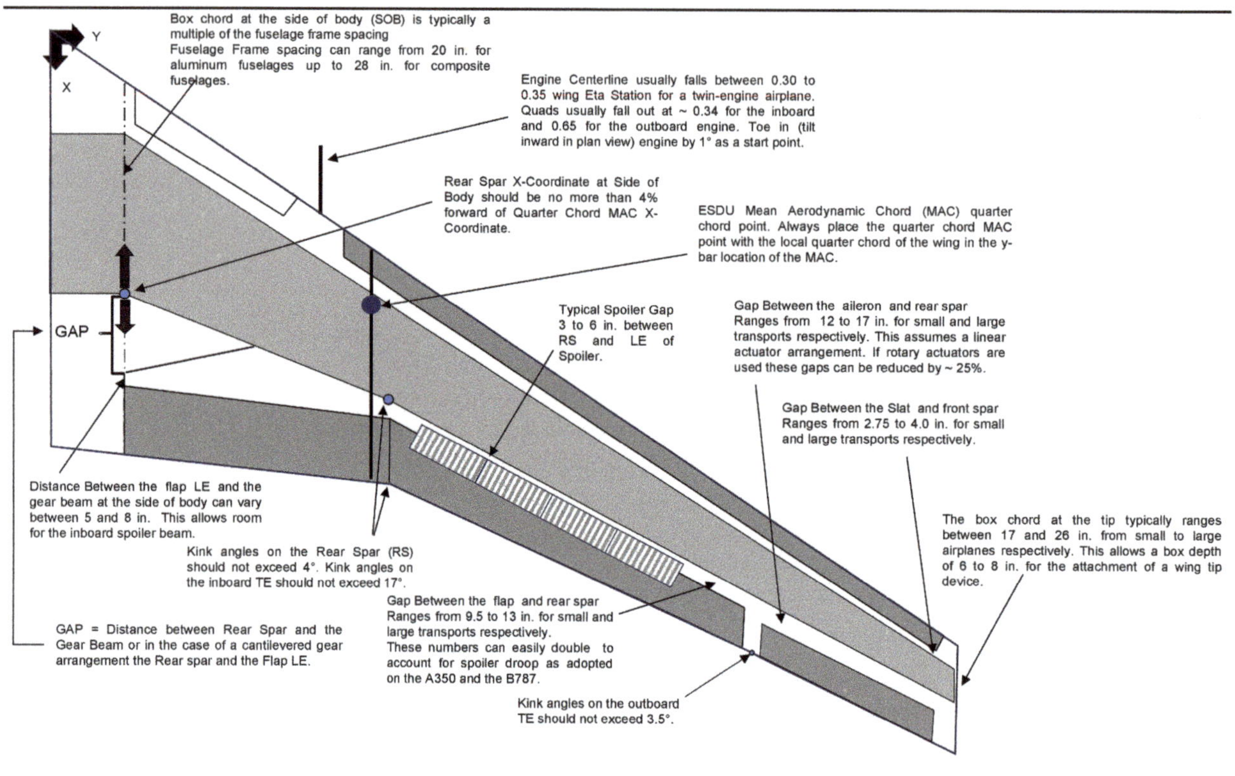

FIGURE 6.5 Gap between the rear spar and the main landing gear beam at the side of body (SOB).

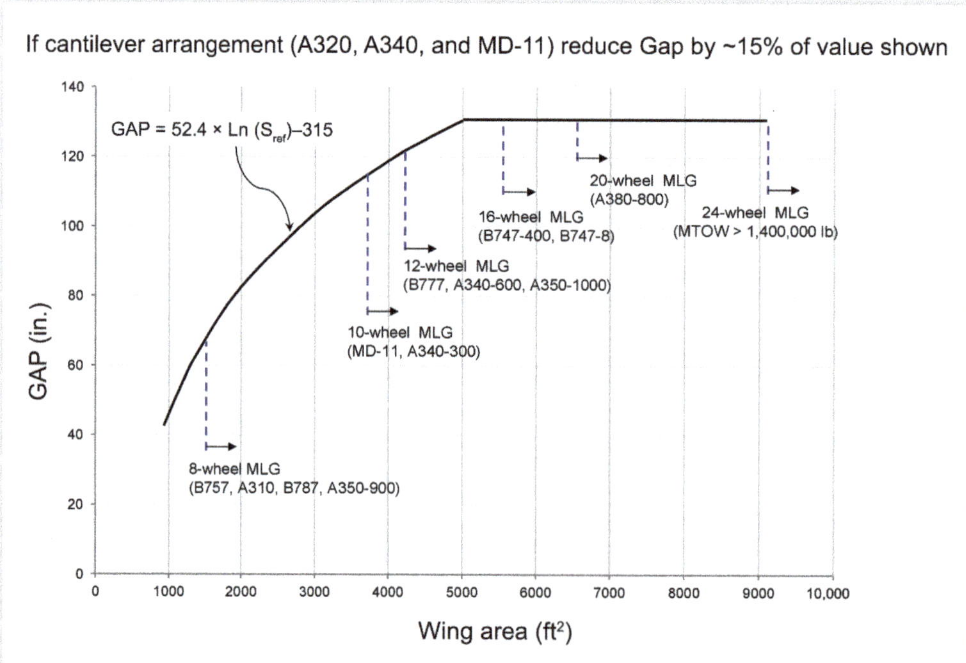

FIGURE 6.6 787 Main landing gear.

Landing Gear and Wheel Well Sizing

Main Landing Gear

The gear length can be approximated by assuming that it will be driven by the nacelle diameter on a low-wing-mounted engine configuration.

$$A = \text{Engine nacelle diameter} + 2.0\,\text{in.} \tag{6.10}$$

$$\text{Trunnion length B} = 0.35 \text{ to } 0.46 \times \text{Dimension} \tag{6.11}$$

Larger (taller) aircraft are closer to 0.35, and single aisles are closer to 0.46.

Note: The trunnion length in this case includes bearings and attach fittings.

Gear position on the wing (y-station from centerline) is calculated as follows:

$$C = 1.16 \times A + 0.0255\text{x } D_f + 10 \tag{6.12}$$

where D_f is the fuselage maximum diameter.

FIGURE 6.7 Main landing gear diagram.

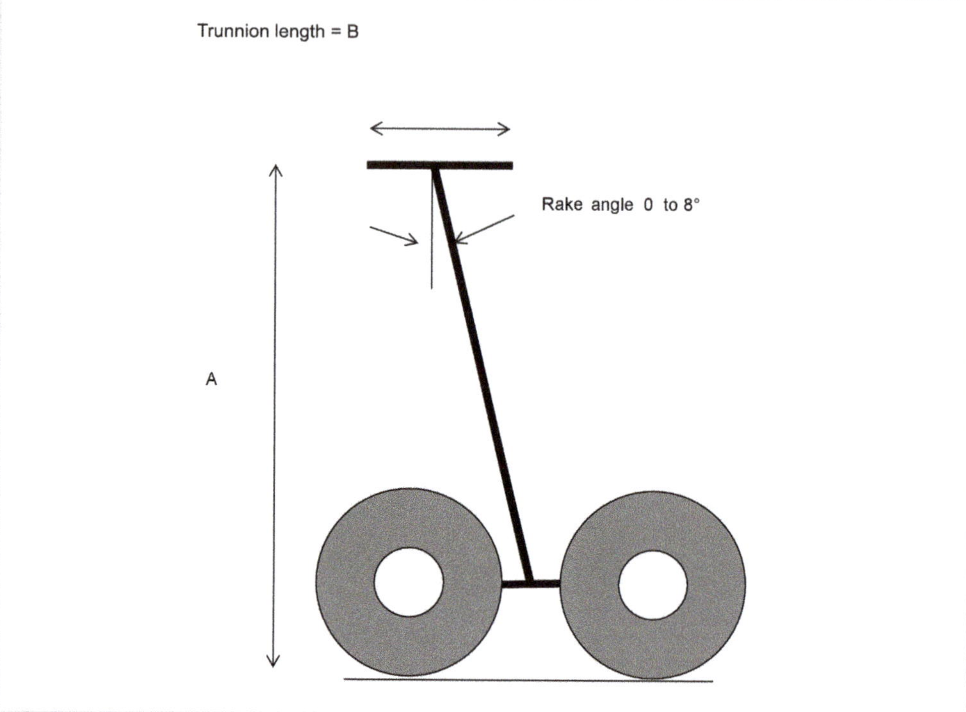

Preliminary Guidelines for Tire Sizing

$$\text{MLG load} = 0.96 \times \text{MTW} \tag{6.13}$$

$$\text{Single-wheel load} = \text{MLG load} \,/\, \text{No. of tires} \tag{6.14}$$

Note: Maximum single-wheel loads should typically not exceed 67,000 lb for an end-of-the-line growth airplane, unless the truck geometry (length and width) is sized to reduce the impact of the aircraft weight on the pavement. High single-wheel loads are an early indicator of impending runway loading issues, which could require airports to strengthen taxiways and runways. For an aircraft family, use the ultimate growth, or largest member of the family to size the wheels and tires, and add an additional 5% to account for uncertainties during development. If the mission requirement for the ultimate growth family member is not well understood in the early phase of design, use the initial member of the family and size the wheels, tires, brakes, and wheel well for an additional 20% growth. This is because landing gear components can be long lead items that cost a significant amount of time and money to develop. Furthermore, tire and wheel commonality between different members of the same aircraft family can reduce the required spares inventory for an airline.

Several tire manufacturers have tire tables and data available online, and this is the preferred method of choosing a tire size. When these data are available, the rated load (static load) can be chosen to match the single-wheel load capability we estimated (including the margin required based on the family member). In the absence of detailed tire tables, the MLG tire size can also initially be chosen using data from existing transports.

While sizing the wheel well (in plan view), adequate room should be allowed for the main landing gear geometry with the oleo or strut in the extended position. Typical main landing gear oleo or strut extension is around 24 in. ± 2 in.

Nose Landing Gear

The nose landing gear length can be drawn using the side view of the aircraft general arrangement drawing. The wheel well geometry and volume should consider the extended oleo or strut of the nose landing gear. The typical nose landing gear extension is around 17 in. ± 2 in.

The aircraft attitude on the ground should typically not exceed 1° nose-down attitude if a freighter version is planned in the future. This is because some freight operators still load cargo manually. Loading and unloading heavy pallets against a slope have ergonomic and safety implications. A slope can also put an excessive burden on a powered cargo handling system, adding weight and complexity to the system design.

Preliminary Guidelines for NLG Tire Sizing

$$\text{NLG tire rated load} = 0.054 \times \text{MTW} \tag{6.15}$$

$$\text{NLG max breaking load} =\sim 1.5 \times \text{NLG tire rated load} \tag{6.16}$$

In the case of the NLG, the single-wheel load is the same as the nose landing gear load. This is because the remaining tire on the NLG must be able to take the load in case the other one goes flat.

Note: The maximum estimated single-wheel load for the nose landing gear is representative of the maximum breaking load. When choosing the tire size, make sure that both load cases (maximum breaking load and rated load) for the tire are considered.

Typical NLG width (tire centerline to tire centerline in front view) varies from 14 to 36 in. for small to large airplanes, respectively.

FIGURE 6.8 Landing gear dimensions of existing aircraft.

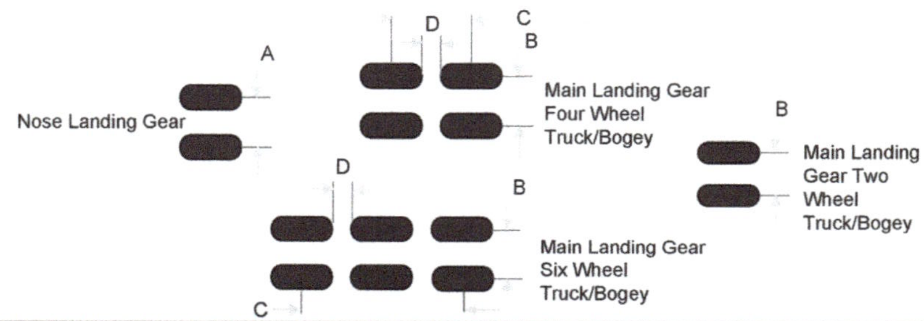

Aircraft	B717-200	A220-300	B737 MAX 8	A321neo XLR	B757-300	B767-300ER	A330-300	B787-9	A350-900	B777-300ER	B747-8	A380-800
MAX TAXI WEIGHT LB. Number of NLG Tires Number of MLG Tires	122,000 2 4	154,500 2 4	181,700 2 4	223,200 2 4	271,000 2 8	413,000 2 8	535,500 2 8	561,500 2 8	619,280 2 8	777,000 2 12	990,000 2 16	1,271,000 2 20
MLG Single Wheel Load lb Tire Speed Rating MPH	29,280 225	37,080 225	43,608 235	53,568 225	32,520 235	49,560 235	64,260 235	67,380 235	74,314 235	62,160 245	59,400 235	61,008 235
DIM (A) inches	14.0	18.57	16.0	19.7	24.0	25	27.95	29	29.45	30.8	36	41.3
DIM (B) inches 2 Wheel / 4 Wheel Truck / 6 Wheel Truck	26 NA NA	35 NA NA	34.0 NA NA	36.5 NA NA	NA 34.0 NA	NA 45 NA	NA 55.0 NA	NA 60.0 NA	NA 68.30 NA	NA NA 55.0	NA 46.8 NA	NA 53.1 60.2
DIM (C) inches 4 Wheel Truck / 6 Wheel Truck	NA NA	NA NA	NA NA	NA NA	45 NA	56 NA	77.99 NA	59.5 NA	80.32 NA	NA 115.2	56.5 NA	66.9 133.8
DIM (D) inches 4 Wheel Truck / 6 Wheel Truck	NA NA	NA NA	NA NA	NA NA	5.0 NA	10.0 NA	24.0 NA	5.5 NA	25.2 NA	NA 5.8 / 6.0	4.5 NA	12.9 12.9
NLG Tire Size (Diameter x Width) inches	26 x 6.6	27 x 8.5	27 x 7.75	30 x 8.8	31 x 13.0	37 x 14.0	40.5 x 15.5	40 x 16	Metric mm 1050 x 395	43 x 17.5	50 x 20	50 x 20
MLG Tire Size (Diameter x Width) inches	41 x 15.0	42 x 15.0	44.5 x 16.5	49 x 18.0	40 x 14.5	46 x 18.0	54 x 21.0	54 x 21.0	Metric mm 1400 x 530	52 x 21.0	52 x 21.0	54 x 21.0

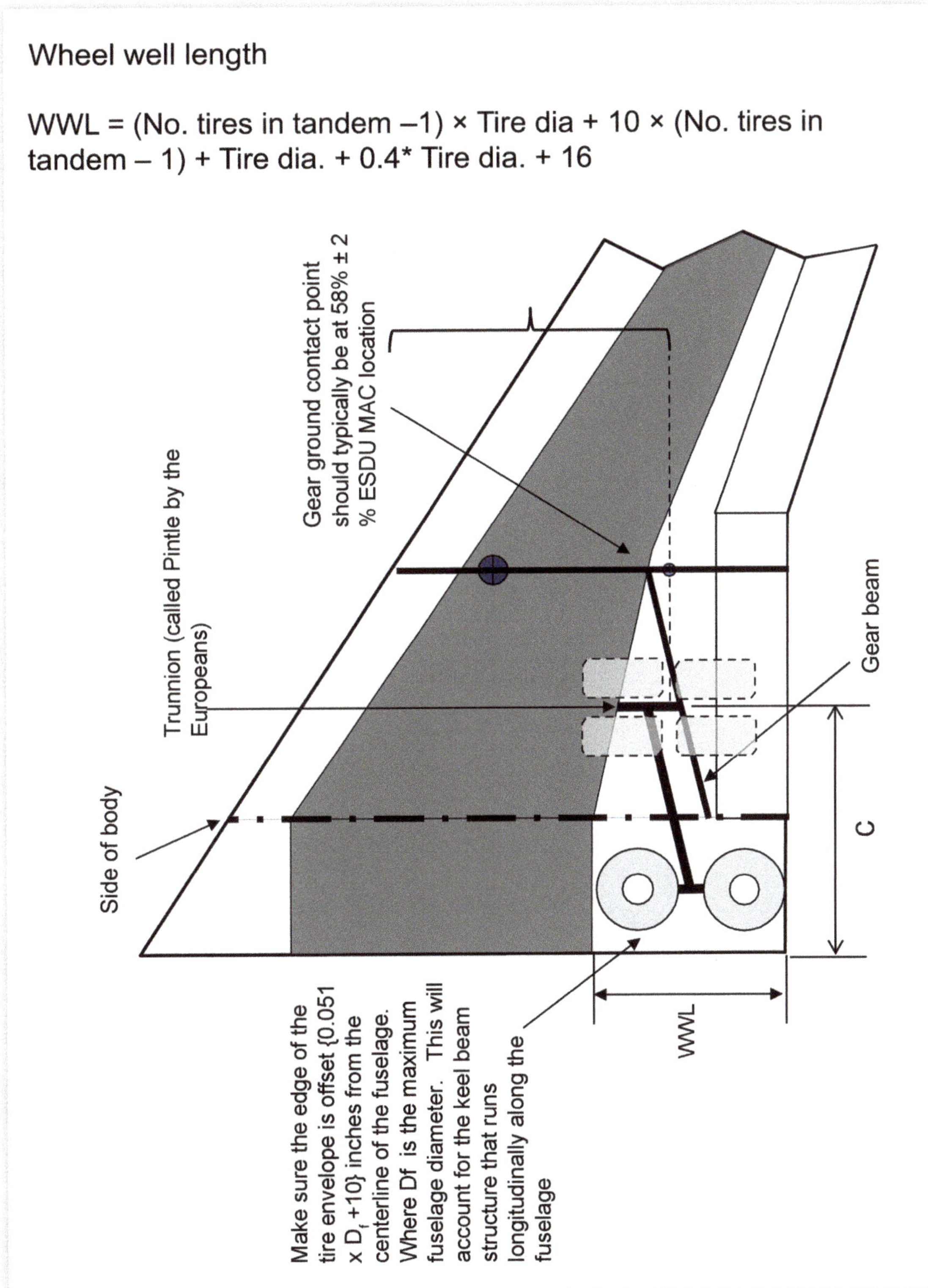

FIGURE 6.9 Inboard wing integration.

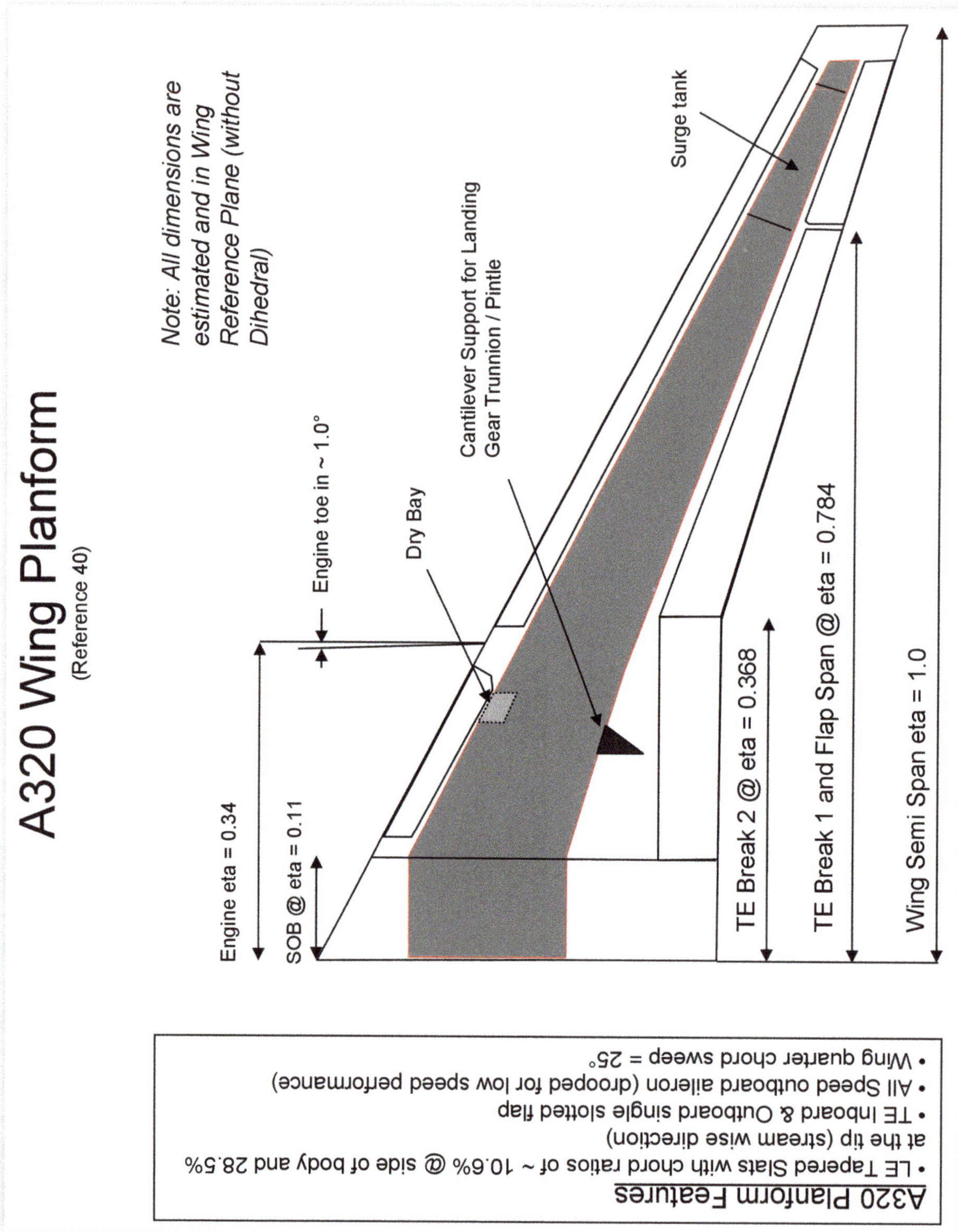

FIGURE 6.10 A320 wing planform.

FIGURE 6.11 A330 wing planform.

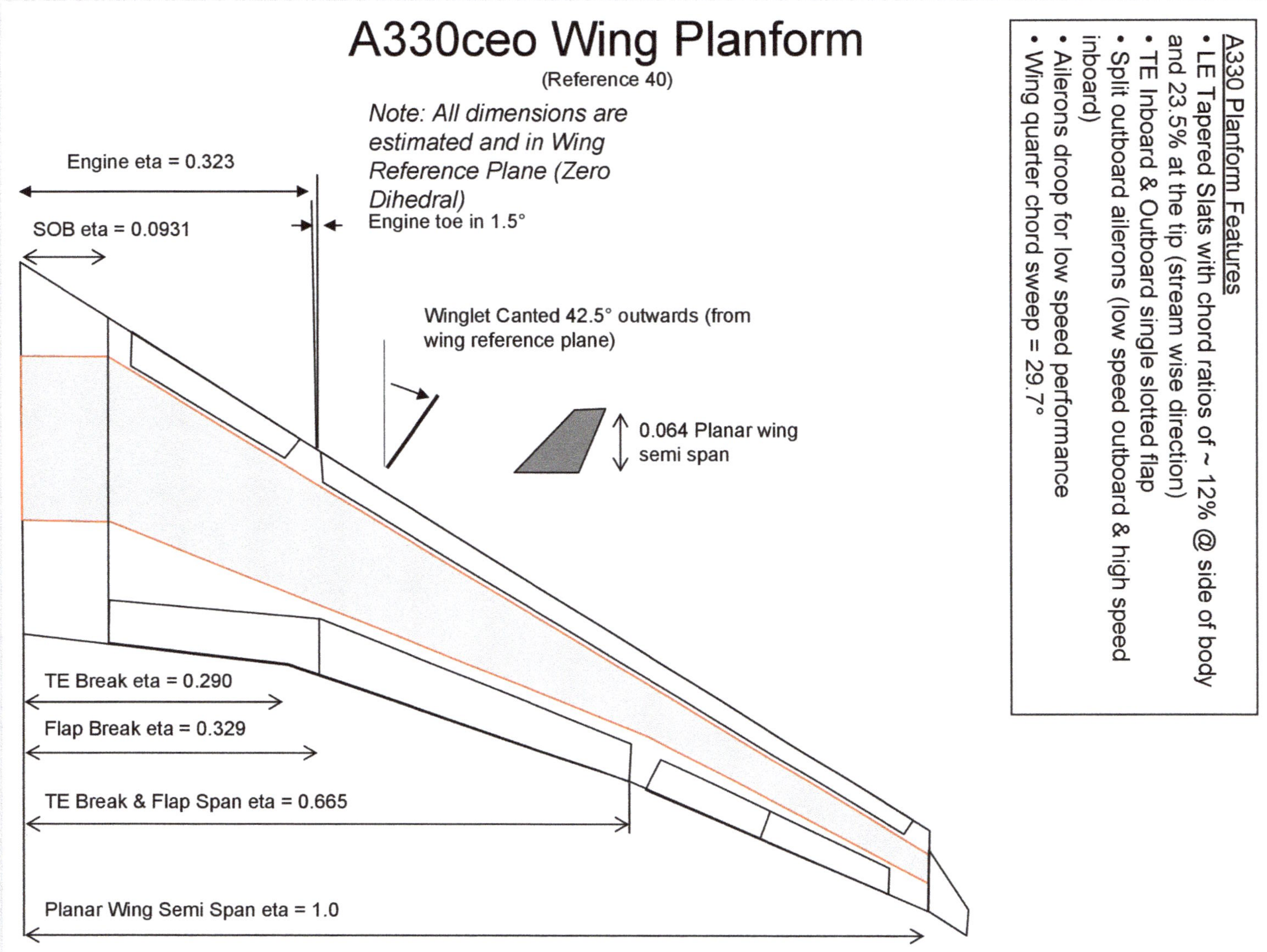

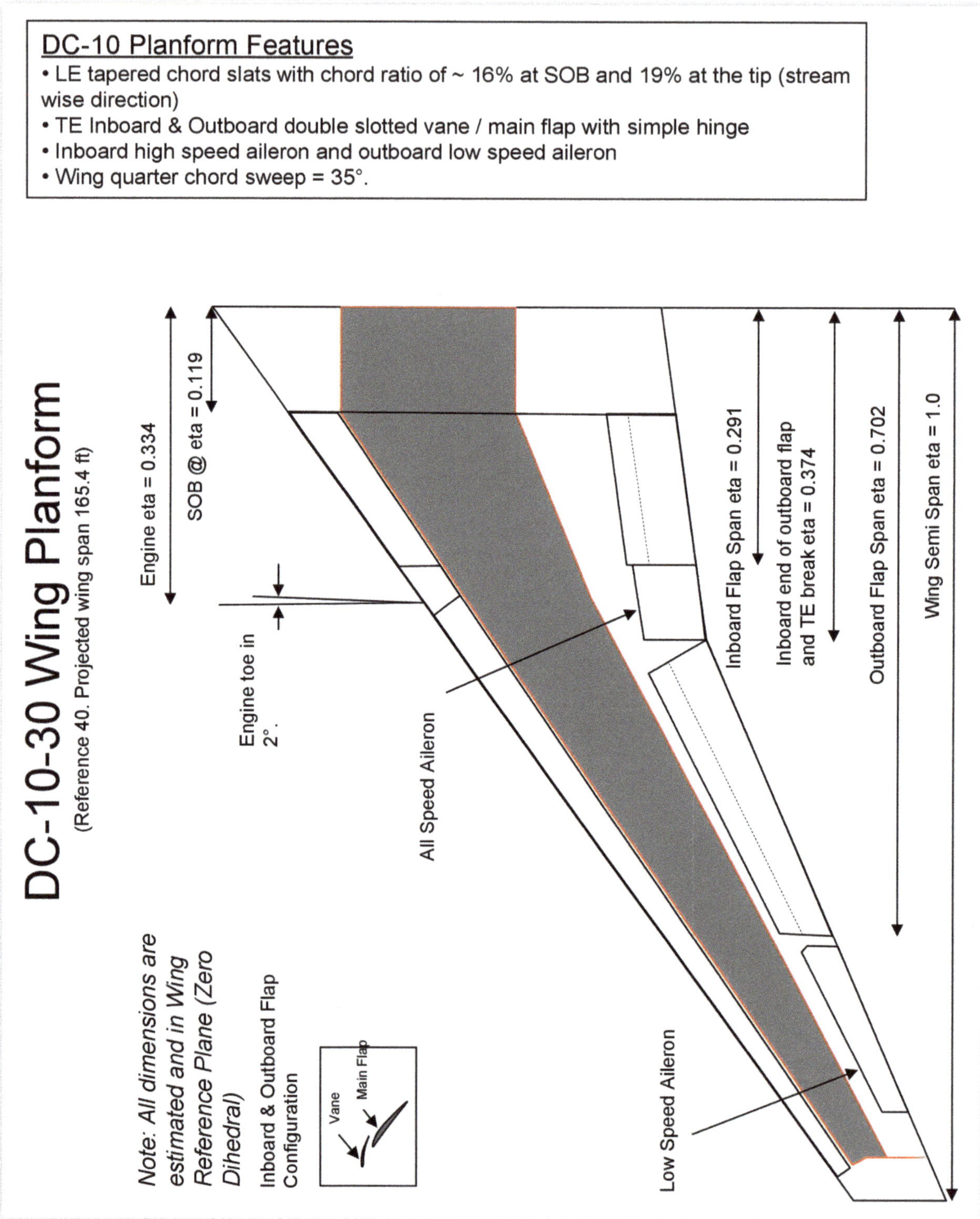

FIGURE 6.12 DC10-30 wing planform.

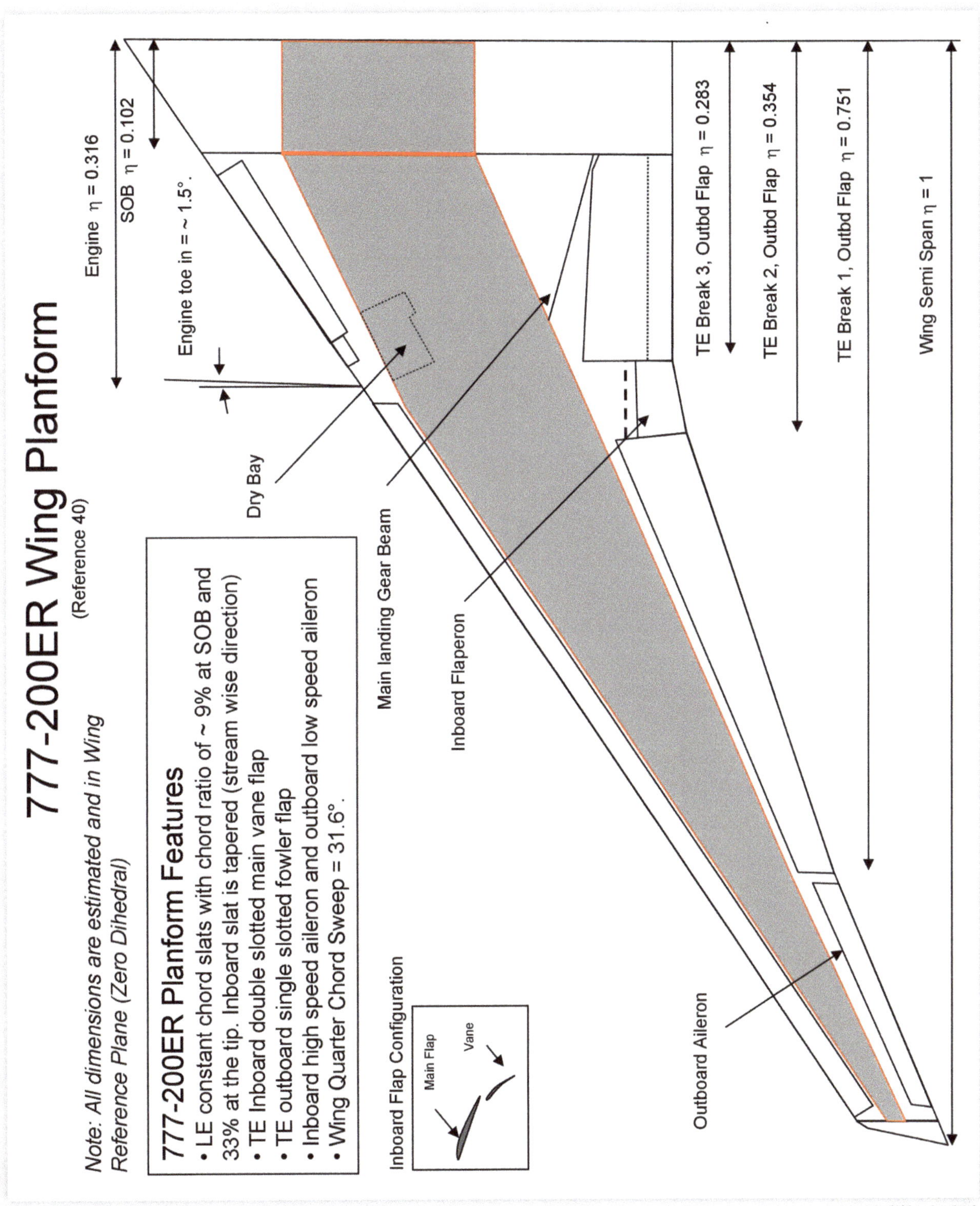

FIGURE 6.13 B777-200ER wing planform.

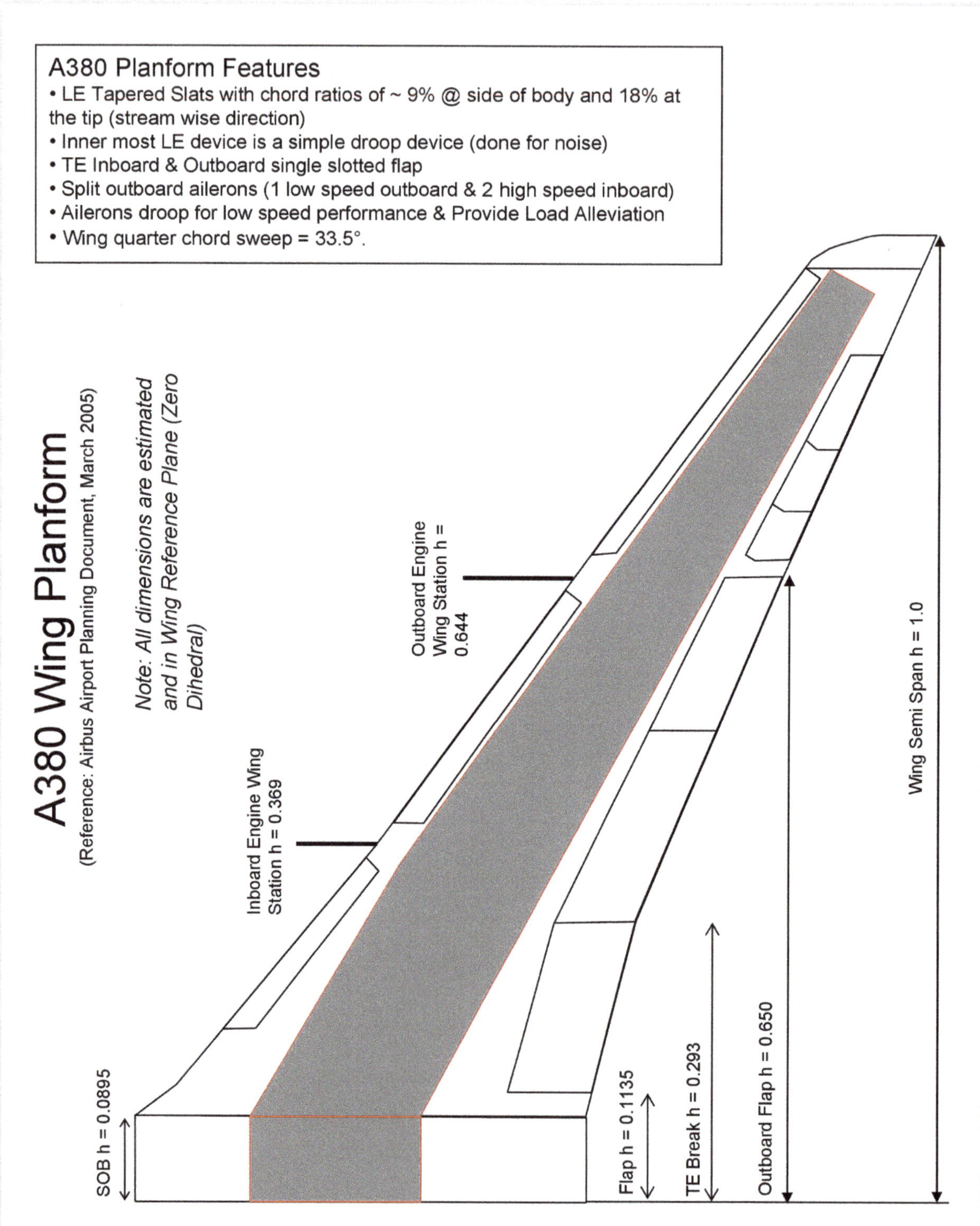

FIGURE 6.14 A380 wing planform.

FIGURE 6.15 Trailing-edge flap chord ratio.

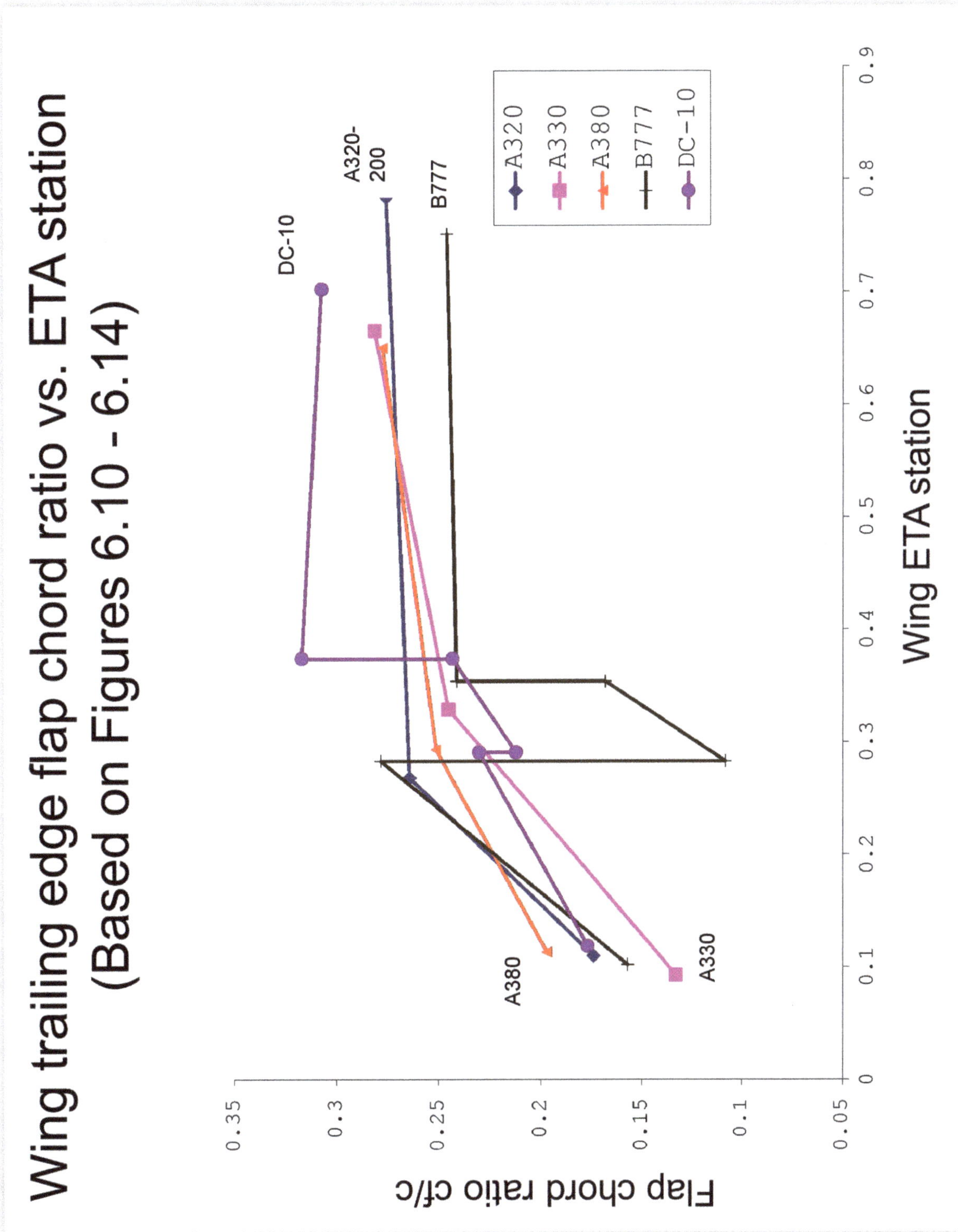

Wing Structure Considerations

In the preliminary design phase, it is important to make sure the wing has the necessary wing structural box area and thickness to handle the torsion and bending loads. As you consider the integration needs of the high lift devices, engine nacelle, and main landing gear, it is also necessary to pay attention to the 2D wing box, specifically the planform area of the exposed wing box from the side of the body to the tip (minus any wingtip device). This is the area available to the wing spars (front and rear), the wing ribs, the wing stringers, and the wing skins.

Providing sufficient wing box area is the first step in ensuring that the structural designers have a good starting point to lay out the spars, the stringers, and the ribs. The figure attached demonstrates how a target wing box planform area can be estimated. One approach this author subscribes to is to optimize the wing planform using the wing box area as a target optimization parameter, while prescribing the space necessary to integrate system components such as actuators.

FIGURE 6.16 Wing box structural area.

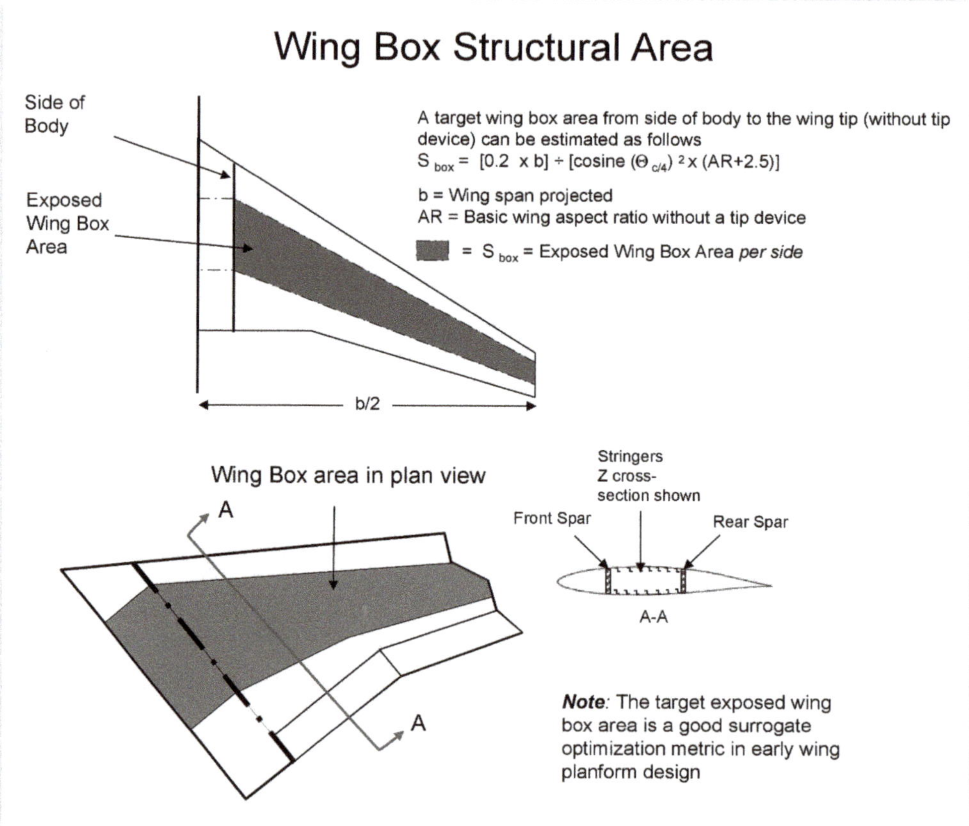

Next, we need to establish a baseline thickness distribution. The t/c (thickness-to-chord ratio) at the side of body (SOB) for most transonic transports ranges from 13 to 16% for cruise Mach numbers from 0.85 to 0.75, respectively.

- The t/c at the tip depends on whether there will be a wingtip device (raked wingtip or winglet). If the wing has a wingtip device, then an absolute thickness of between 6 to 11 in. is desired depending on the length of the tip device. This is to ensure there is enough structural depth to not only hold on to the tip device, but also provide room for a mechanic to install the device. In the planform design process, the cruise Mach number and the need (or not) for a tip device necessitate that a minimum wing planform tip chord be specified to avoid a larger than necessary t/c ratio at the tip, causing an increase in wing drag.
- Once the tip and the SOB t/c are picked, the average outboard t/c (ETA stations 0.35 to 0.95) is driven by the t/c value at some mid-wing stations. The spanwise location of the inboard trailing-edge kink of the wing planform usually works well as a starting point. The t/c at this wing station is varied until the required average "high-speed" outboard t/c is achieved.
- The thickness at the wing inboard trailing-edge kink (ETA between 0.35 and 0.45) sets the thickness to integrate the engine. While side of body wing end loads can be the highest, they compete for that title, with the end loads just outboard of the engine. This is also a sensitive area for high-speed drag, and hence, the choice of t/c creates a tension between needing enough thickness to keep loads down and hang on to the engine, while also allowing for a low-drag wing.
- Lastly, the t/c at this location (ETA 0.35 to 0.45) can drive the curvature of the lower wing skin panel. A "tight" or lower radius of curvature between the side of the body and the engine can lead to producibility challenges, making the wing difficult to build at a high production rate. A high production rate is likely a key requirement to satisfy the high number of units required for the single-aisle market.

FIGURE 6.17 Wing thickness-to-chord distribution.

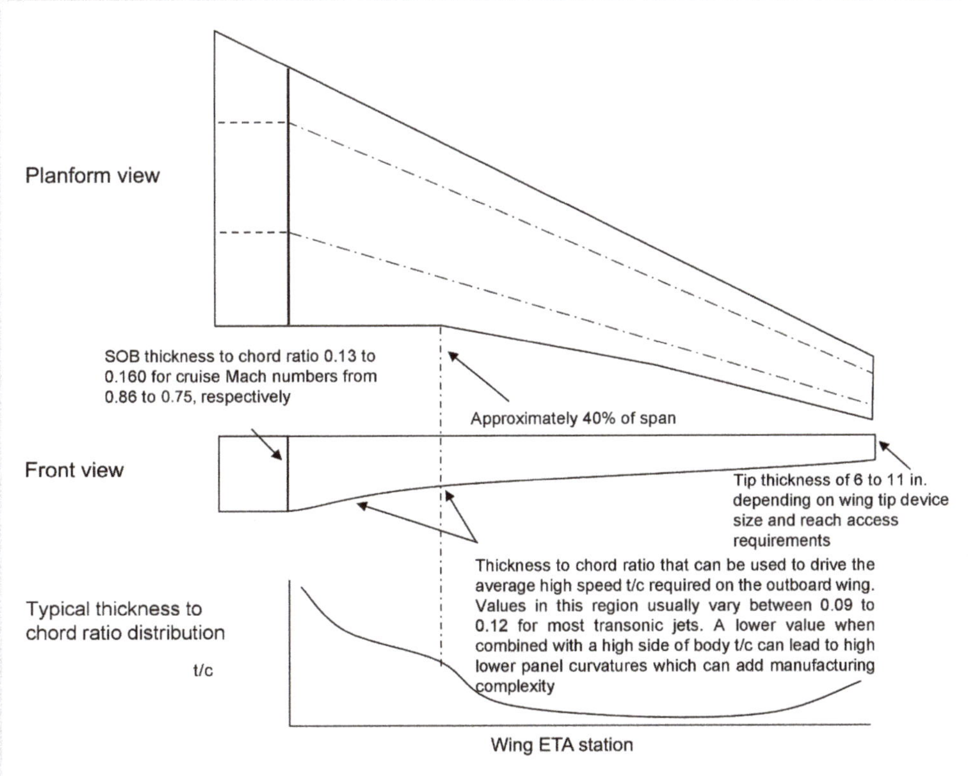

Torenbeek [6] recommends a formula that allows a reasonable side of body thickness to be picked. Along with the wing box area target, this should allow for a reasonable wing planform for a weight-efficient design.

FIGURE 6.18 Wing cantilever ratio.[1]

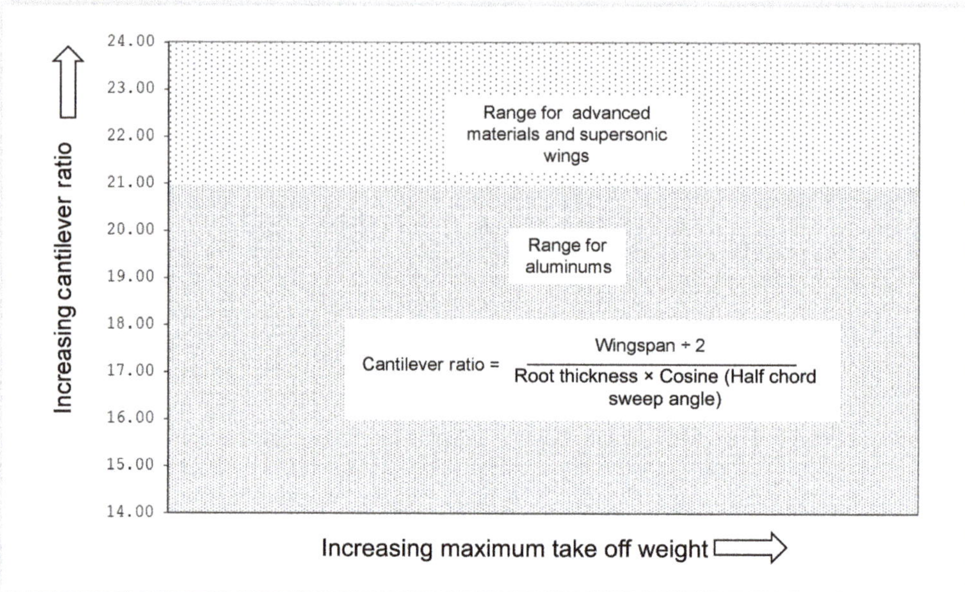

[1] Torenbeek, Egbert, *Synthesis of Subsonic Airplane Design: An Introduction to the Preliminary Design of Subsonic General Aviation and Transport Aircraft, with Emphasis on Layout, Aerodynamic Design, Propulsion and Performance*, Springer Science & Business Media, 2013.

Paying Attention to the Aerodynamics

- Much like in the structure section, the main goal here is to address the wing integration challenges while making sure a robust wing planform exists for aerodynamic optimization. While this approach may seem counterintuitive at first to the aerodynamicist, it is my experience that the product achieved via this approach is a lot more robust both structurally and aerodynamically.
- To achieve a successful design, it is important to achieve a well-integrated wing, while also making sure no span load "kinks" are introduced.
- Much like the structural layout of the spars, etc., we need to ensure that leading-edge and trailing-edge kinks (or angles) flow smoothly when they transition from one kink to the other. On the inboard wing (less than 45% of the semi-span) changes in trailing-edge angles in excess of 17° should be avoided. On the outboard wing (more than 45% in wing semi-span) changes in trailing-edge angles in excess of 3.5° should be avoided.
- Approximating a span load will help us achieve a wing, which is in the ballpark of where it needs to be aerodynamically. Once again, the aim here is not to do a detailed aerodynamic design but to provide the aerodynamicist with a good starting point for aerodynamic optimization.

The Process Steps

1. Use the wing planform drawing to calculate the local chords c(y). The actual wing planform drawing should be used for reference. Here, "y" is the distance of the local chord from the wing centerline.

$$y_1 \le c(y) \le y_0$$

$$c(y) = x_0 + x_1 + y \times [\tan(\theta_{TE}) - \tan(\theta_{LE})] \quad (6.17)$$

$$y_2 \le c(y) \le y_1$$

$$c(y) = x_0 + x_1 + x_2 + y \times [\tan(\theta_1) - \tan(\theta_{LE})] \quad (6.18)$$

$$y_3 \le c(y) \le y_2$$

$$c(y) = x_0 + x_1 + x_2 + x_3 + y \times [\tan(\theta_2) - \tan(\theta_{LE})] \quad (6.19)$$

$$y_{SOB} \le c(y) \le y_3$$

$$c(y) = x_0 + x_1 + x_2 + x_3 + x_4 + y \times [\tan(\theta_3) - \tan(\theta_{LE})] \quad (6.20)$$

   Knowing the local chords allows us to get a first estimate of the required local airfoil section lift coefficients or cls. These data can then be utilized to derive airfoil sections that can meet the required cls.

2. To derive a plot of the local cl distribution, we also need to assume a span load. The chart that follows shows a derived span load of a wing drawn by the author. The assumed span load combined with the local chord distribution along the wingspan is used to calculate the first estimate of the local section lift coefficient.
3. A parameter defined as c/c_{bar} is used to normalize the chord distribution with span. Here, "c" is the local chord at any given station (derived from the planform) and "c_{bar}" is the average wing chord defined previously as the wing area divided by the wingspan (without tip device). Applying the c/c_{bar} of a new wing along with the design C_L from the sizing exercise, a local cl plot for the new wing can be derived.

FIGURE 6.19 Span load versus wing ETA station.

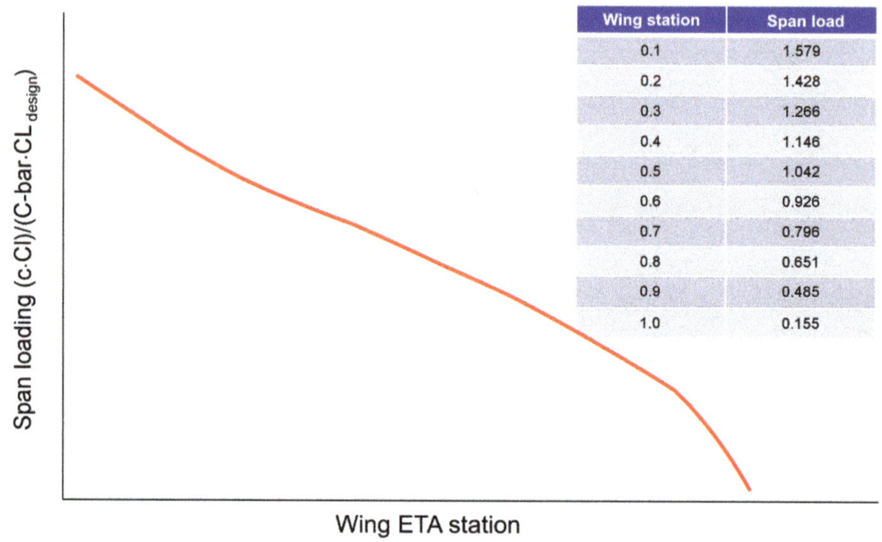

Wing station	Span load
0.1	1.579
0.2	1.428
0.3	1.266
0.4	1.146
0.5	1.042
0.6	0.926
0.7	0.796
0.8	0.651
0.9	0.485
1.0	0.155

To derive a plot of the local Cl distribution we will assume a span load, shown in the chart above. Applying the c/c-bar of a new wing along with the design C_L from the sizing exercise a local Cl plot for the new wing can be derived.

4. The chart below shows the c/c_{bar} distribution of a wing for a four-engine transport and a twin-engine transport (which was used to derive the span load on the previous chart). Both have been derived by the author. As one can see, the four-engine transport has more chords on the outboard wing.

FIGURE 6.20 c/c_{bar} versus wing ETA station.

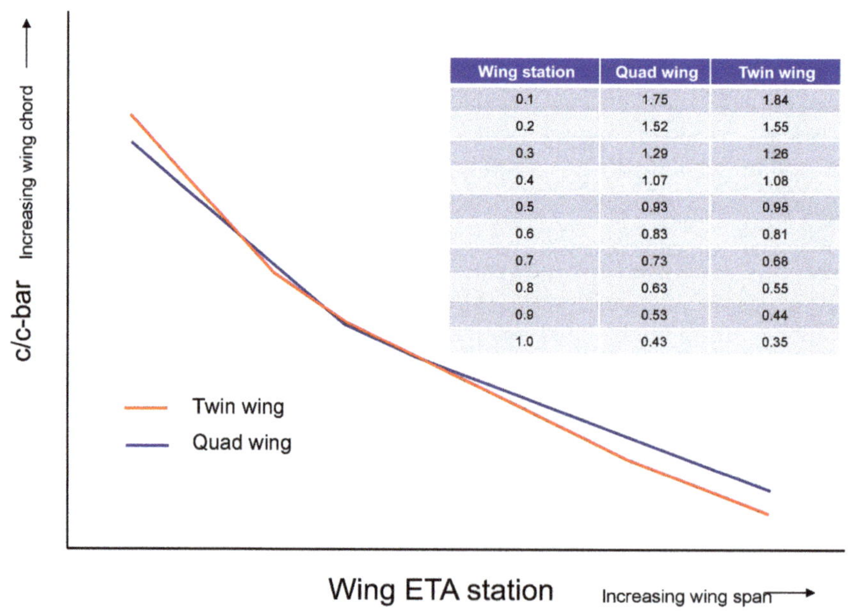

Wing station	Quad wing	Twin wing
0.1	1.75	1.84
0.2	1.52	1.55
0.3	1.29	1.26
0.4	1.07	1.08
0.5	0.93	0.95
0.6	0.83	0.81
0.7	0.73	0.68
0.8	0.63	0.55
0.9	0.53	0.44
1.0	0.43	0.35

5. Primarily, this is due to the four-engine transport requiring more depth in the outboard wing to mount the outboard engine, along with the fact that four-engine transports usually tend to have a lower aspect ratio due to the less critical second segment climb requirement—this allows for longer chords on the outboard wing.

FIGURE 6.21 Estimating the wing span load.

The local lift coefficient in Col (4) can be calculated at each ETA station as follows:

Col (4) = Col (2) + Col (3) x Design C_L

Derived from new wing → Col (3); Calculated for new wing → Col (4)

Col (1)	Col (2)	Col (3)	Col (4)
Wing ETA station	Base span load	c/c_{bar} New wing	Local Cl New wing
Eta side of body	1.579		
0.2	1.428		
0.3	1.266		
0.4	1.146		
0.5	1.042		
0.6	0.926		
0.7	0.796		
0.8	0.651		
0.9	0.485		
1.0	0.155		

6. Applying a twist to the wing is a sophisticated process, which must consider the influence of the engine nacelle, while also seeking the optimum balance between wing weight, induced drag, and wave drag. At this stage of the design, we are more interested in the geometric twist for the purpose of approximating the geometry to install the engine, such that we have adequate ground clearance. This can be an iterative part of the design process, and the aim is to provide a stable wing, engine, and landing gear geometry for further analysis.

   The total wing twist can be approximated as follows:

$$\text{Total twist} = 2.6 + \left\{CL_{\text{cruise}} \times 1.03 \div \text{Wing} - \text{body lift curve slope}\right\} \tag{6.21}$$

7. Here, CL_{cruise} is the mid-cruise trimmed design lift coefficient multiplied by a factor to account for tail download. A value of ~3% can be used to approximate the download of the tail. If fuselage fineness ratios are much lower than 10, then a 5% factor is more appropriate. This is because the wing must carry additional lift to account for the tail, which is lifting downward.

8. The wing is twisted such that it is mounted on the fuselage at some angle of incidence. The wing incidence on the fuselage is typically driven by the top-of-climb (TOC) condition. A fuselage angle of attack of no more than 2.5° is usually desired to enable cabin crew to start meal services immediately after reaching cruise altitude. If the fuselage angle of attack is much over 2.5°, cabin crew find it difficult to push a fully loaded galley cart from the aft galley and up the aisle. Most commercial transports are designed to a fuselage angle of attack criteria at TOC of 3° ± 0.5°. Increased wing incidence angle on the fuselage comes with an increased drag penalty.

The airplane's TOC lift coefficient can be got from our sizing and is approximately equal to the maximum usable CL of the aircraft. In our case, this value is $CL_{TOC} = CL_{usable}$

Note: This value as before will be multiplied by 3% to account for the tail download.

$$\text{Wing incidence} = \left\{CL_{usable} \times 1.03 \div \text{Wing-body lift curve slope}\right\} - \text{Fuselage angle of attack}(\text{TOC}) + 1.5° \quad (6.22)$$

FIGURE 6.22 Wing incidence on an A340-300.

Assuming a uniform twist distribution, we can estimate the geometric twist of the wing from root to tip.

$$\text{The geometric twist} = \text{Total twist} - \text{Aerodynamic twist} \quad (6.23)$$

For preliminary design, we can assume the aerodynamic twist to be approximately 1°.

Note: Washout is the condition where the tip of the wing is at a lower angle of attack than the root. Washout typically means a negative twist angle.

For example:

$$\text{If} \quad \text{Total twist} = 7^\circ = -7^\circ \quad \text{for washout}$$

$$\text{Aerodynamic twist} = 1^\circ = -1^\circ \quad \text{washout}$$

$$\text{Geometric twist} = \text{Total twist} - \text{Aerodynamic twist}$$

$$= -7 - (-1) = -6^\circ$$

The three-dimensional wing can now be defined given the wing planform, thickness distribution, and twist. This together with the landing gear can be used to position the wing on the fuselage and then install the engines and add the vertical and horizontal tails.

FIGURE 6.23 Estimating the wing spanwise geometric twist.

Note: The incidence angle is applied at the side of body. The same incidence will run through the wing center section. Outboard of the side of body we will apply the above formula to get the geometric twist of each wing section

Col (1)	Col (2)	Col (3)	Col (4)	Col (5)
Wing ETA station	Wing incidence I at SOB (degrees)	Side of body chord (in.)	Local chord (in.)	Airfoil geometric incidence angle (degrees)
D_f/b	I = 2.95			2.95
0.2				
0.3				
0.4				
0.5				
0.6				
0.7				
0.8				
0.9				
1.0				

$$\text{Col (5)} = \text{I} - \text{Absolute value} \left\{ \frac{\text{Geometric twist} \times [\,1 - \text{Square root}\,(1 - \text{Col (1)})\,]}{(\text{Col (3)} \div \text{Col (4)})} \right\}$$

Note: Accounting for the twist distribution at this early design phase allows for a reasonable chance of integrating the engine relative to the ground plane.

FIGURE 6.24 Geometric twist on jet transports of a different vintage.

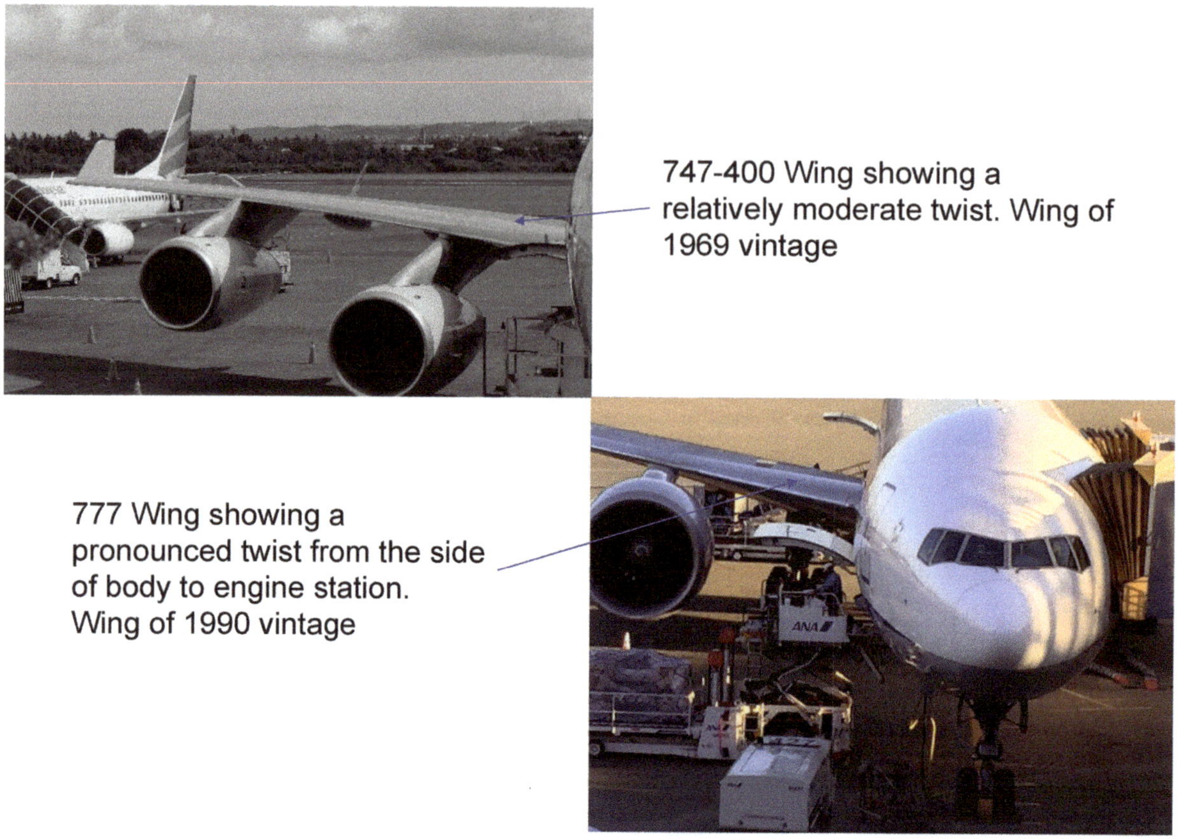

Cavan-Images/Shuttersto

7

Layout of the Horizontal Tail

Definitions

Tail Volume Coefficient

The tail volume coefficient (V_h) is a measure of the effectiveness of the horizontal stabilizer relative to the wing. The tail volume coefficient reflects the contribution of the tail to the longitudinal stability of the aircraft.

The shortest member of the family usually sets the area of the HT, and in most cases, a manufacturer will usually introduce the smallest member of the family first to realize the efficiencies associated with the heavier stretch models. However, this is not always true, such as in the case of the Boeing 777X family where Boeing has chosen to introduce the B777-9 first, followed by the shorter B777-8. The 777-8 will set the HT area for the remaining family members.

The tail volume coefficient is calculated as follows:

$$V_h = (l_h \times S_h) / (S_{ref} \times c_w) \tag{7.1}$$

where
l_h is the distance of the HT quarter-chord MAC to the center of gravity of the aircraft. For practical purposes in the early design phase, the quarter-chord of the wing MAC is used instead of the center of gravity.
S_h is the horizontal tail area
S_{ref} is the wing area
c_w is the mean aerodynamic chord of the wing

However, since little is known about the wing planform, we can use the average mean chord instead. The average mean chord is simply the wing area divided by the wingspan.

$$c_w = S_{ref} / \text{Wingspan} \tag{7.2}$$

Empirically, the volume coefficient can also be related to some of the known fuselage and wing parameters such as wing area, the mean geometric chord of the wing, the fuselage length, and fuselage width.

$$V_h = 0.365 \times K_f + 0.67 \tag{7.3}$$

where

$$K_f = \left[W_f^2 \times L_f\right] / \left[S_{ref} \times c_w\right] \tag{7.4}$$

By equating the two equations for the HT volume coefficient, we can solve for the HT area.

$$S_h = \left\{0.365 \times \left[W_f^2 \times L_f\right] / l_h + 0.67 \times \left[S_{ref} \times c_w\right] / l_h\right\} \times k \tag{7.5}$$

where
L_f is the fuselage length (ft)
W_f is the fuselage width (ft)
k = 1.0 for HT without active CG control trim tank (with or without HT fuel tank)
= 0.9 for HT with an active CG control trim tank

In the case of a new project aircraft, we can estimate certain dimensions, which can be refined later as more aircraft configuration design details become available.

- $l_h = 0.510 \times L_f$ Conventional configuration with T-tail (BAE 146)
- $l_h = 0.455 \times L_f$ Conventional configuration with low tail (B777, A380, etc.)
- $l_h = 0.480 \times L_f$ Aft engine configuration with T-tail (Fokker F70)
- $l_h = 0.430 \times L_f$ Aft engine configuration with low conventional tail (MD-11)

Note: The wingspan should be the effective aerodynamic wingspan.

FIGURE 7.1 Typical pitching moment diagram.

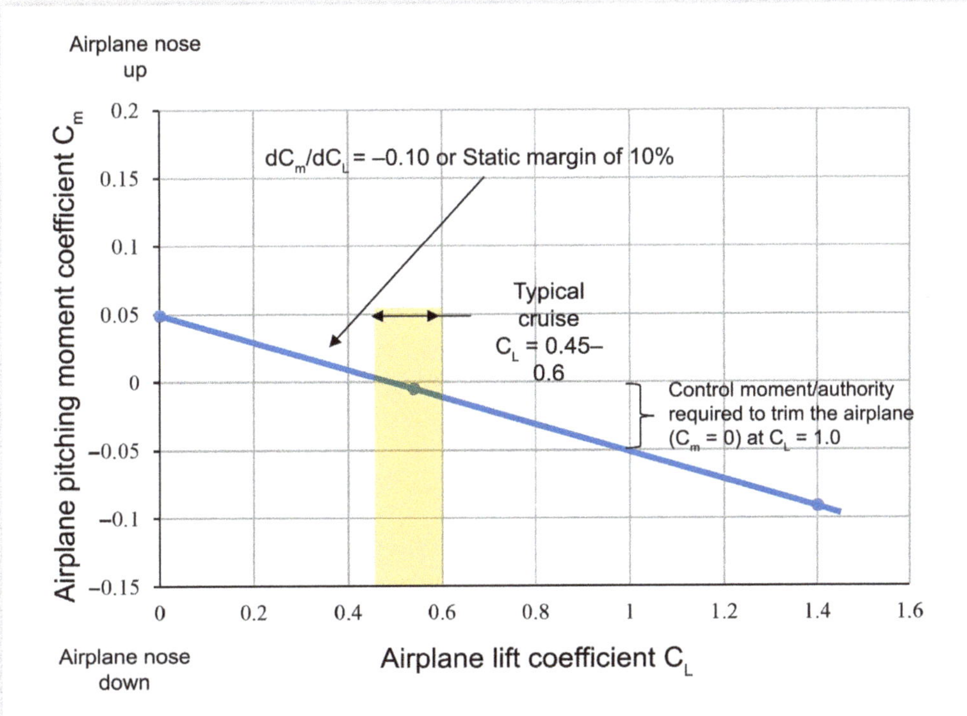

Reference Horizontal Tail Information

Reference [29] has the horizontal tail areas of current transports (Table 7.1). It also serves as a good check for a particular aircraft type and/or technology level. The tail volume coefficients are calculated using the same approximations mentioned earlier, since accurate tail arm data, etc., is proprietary in nature.

TABLE 7.1 Horizontal tail reference information.

Aircraft type	Wing area S_{ref} (ft²)	HT area S_h (ft²)	Ratio S_h/S_{ref}	Comments
A320neo	1310	334	0.252	Fly-by-wire (FBW) with relaxed static stability (RSS)
Fokker 70	1006	234	0.232	Conventional stability T-tail aircraft with aft-mounted engines
A340-500	4704	1001	0.213	FBW RSS. Horizontal tail fuel for trimming with four engines
A350-900	4650	890	0.191	FBW RSS
B777-200ER	4605	1090	0.236	Fly-by-wire RSS
MD-11	3648	920	0.252	Tri-jet with FBW and RSS and HT fuel tank for trimming. Elevators are slotted for the improved control authority

Aspect Ratio

Horizontal tail aspect ratios range from 4.5 to 5.5. The higher aspect ratio makes for a more aerodynamically effective, but structurally heavier tail. Aerodynamic efficiency is realized via an increase in the lift curve slope or CL_α. Hence, for a given angle of attack the HT will be more powerful. This helps with reducing cruise trim drag and increases control power. For the same control power, the higher aspect ratio allows the horizontal tail area to be reduced, which reduces both weight and profile drag. This decrease in area is typically balanced against increased structural weight for the more slender surface.

Taper Ratio

Horizontal tail taper ratios range from 0.25 to 0.35. The higher taper ratio can also make for a more efficient but potentially heavier empennage. A higher taper ratio helps to improve the induced drag of the horizontal tail. A more efficient HT in cruise helps reduce the aircraft's overall trim drag. This decrease in trim drag is typically balanced against increased structural weight for a potentially more slender box structure at the root. In the context of horizontal tail layout, the taper ratio is often used to help integrate the different components such as the elevator actuation system and an adequate box chord.

Thickness-to-Chord Ratio and Quarter-Chord Sweep

Thickness-to-chord ratios typically range from 9 to 10%. The quarter-chord sweep of the horizontal tail is typically set anywhere between 0 and 3° higher than the quarter-chord sweep of the wing. The aim here is to make sure that the combination of sweep and thickness allows the HT to maintain its control effectiveness long after the wing is seeing shock-induced separated flow. In other words, you want to make sure that the HT can help pull the airplane out of a dive at the maximum dive speed and dive Mach number.

Dihedral

Conventional jet transports usually carry a fair amount of dihedral on the horizontal tail. This is usually done to make sure that with the flaps deployed and in certain high-speed maneuvers, the disturbed or separated flow coming off the wings does not cause the HT to vibrate. The typical dihedral angle for a low-mounted HT (B777, A330, and MD-11) is 7°, while that of a T-tail configuration (CRJ-700 and Fokker 70) will be 0° or even −1° (anhedral).

Center of Gravity

The center of gravity (CG) of an aircraft is usually defined in terms of the mean aerodynamic chord of the wing. The mean aerodynamic chord (MAC) of a given wing is the width of an equivalent rectangular wing in each set of aerodynamic conditions. The CG range (forward to aft) in terms of the MAC is set by the length of the fuselage (a longer fuselage means a longer cabin) to account for variations in passenger and cargo loadings. The horizontal tail can be sized by either take-off or landing in forward CG conditions both of which require enough tail area for trim with some additional capability for control.

The aft CG may also drive tail size due to a "push over" control input necessary during an aborted landing. In the landing configuration, the HT is usually trimmed in a nose-up position to cancel the pitch-down moment of the deployed flaps. In a go-around situation, the pilot usually pushes the throttles to maximum power and then retracts the flaps to a detent optimized for an efficient climb out. Both actions can cause the aircraft to pitch up, and enough control authority must be available from the HT to force the nose down to the appropriate pitch angle for the climb. If the aircraft is in an aft CG condition, the pitch-up tendency is exacerbated. Hence, depending on the jet transport configuration both the forward and aft CG conditions can drive the size of the horizontal tail.

Ice Protection

One of the critical criteria that can size the horizontal tail is the forward CG condition in the landing or take-off configuration. If the leading edge of the horizontal tail does not have an ice protection system or if the aerodynamic geometry (airfoil leading-edge shape, sweep, and thickness) of the horizontal tail is more conducive to ice formation, then it may be necessary to increase the area of the horizontal tail. This can affect both the weight and drag of the aircraft. Utilizing an ice protection system among other aerodynamic improvements (active HT fuel trim tank) allowed the MD-11 to have a 30% reduction in tail size over the DC-10.

Relaxed Stability

A longitudinally stable aircraft is one in which the aircraft has the natural tendency to return to its initial trimmed flight path when disturbed. Older generation jet transports had their aft center of gravity limit, forward of the aircraft's neutral point by 10% of the mean aerodynamic chord. The advantage of moving the CG further aft is that it allows you to reduce the trim drag of the airplane. This, however, means that the aircraft will be more sensitive in the pitch axis and may require an advanced flight control system to assist the pilot with control. The aft limit is usually set by considerations of a flight control computer system failure, which requires that the pilot be able to fly the aircraft in direct mode (minus the assistance of the digital flight computers). Relaxed stability can be different for different aircraft and can range from +5% MAC (CG ahead of the neutral point) to −5% MAC (CG aft of the neutral point).

Horizontal Tail Fuel Tank

Putting a fuel tank in the horizontal tail can serve several purposes. Tail fuel can be a consideration during the early sizing phases of the design process. If the wing is sized by fuel volume constraints, it may be advantageous to consider a tail tank to store additional fuel and reduce the overall size of the wing. This can reduce both weight and drag. Making the tail tank an active part of the fuel management system can help move the CG aft during the cruise phase of flight, thereby reducing trim drag. This can help further size the tail area down because the forward and aft CG can also be more actively managed to meet the take-off and landing control requirements. However, putting a fuel tank in the tail is not trivial. It requires that the tail now have full lightning protection with the added complexity (weight and maintenance cost) of fuel pumps, fuel transport elements, and a stronger structure to accommodate the additional loads. The decision to incorporate a tail fuel tank can vary from aircraft to aircraft depending on the configuration's individual characteristics. For example, the MD-11, A330, A340, and A380, all have an active tail tank, while the newly proposed A350XWB does not have one. In the Boeing fleet, the B747-400 has a passive tail fuel tank, which is not used for active CG management. This was primarily done to increase the fuel volume capacity of the aircraft. Adding an actively controlled tail fuel tank will typically reduce the horizontal tail area by ~10%.

FIGURE 7.2 Horizontal tail general definition.

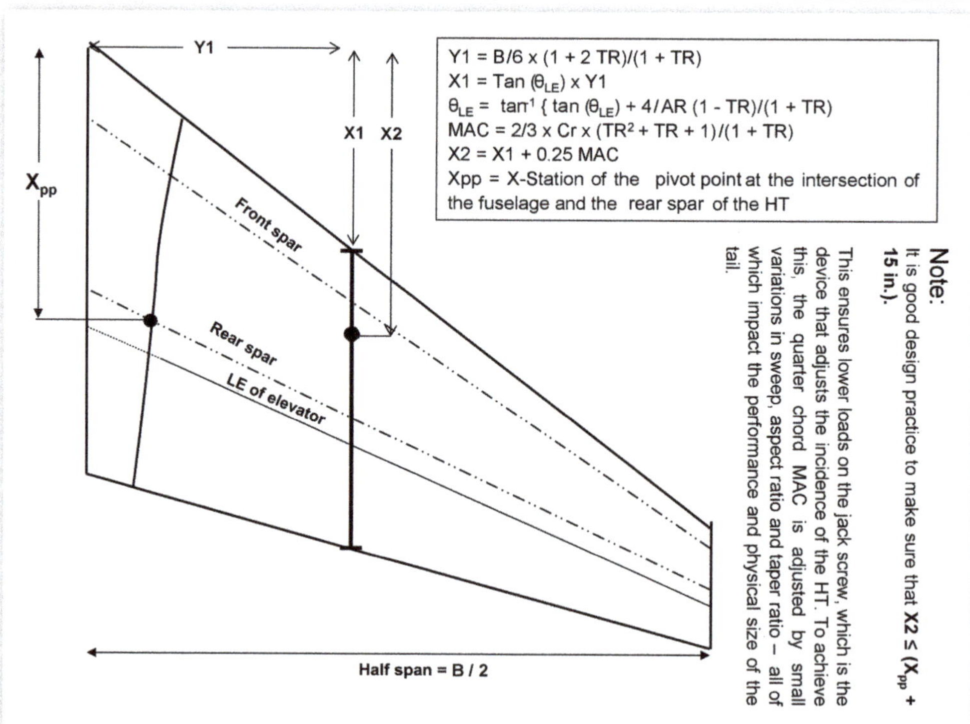

FIGURE 7.3 Horizontal tail layout guidelines.

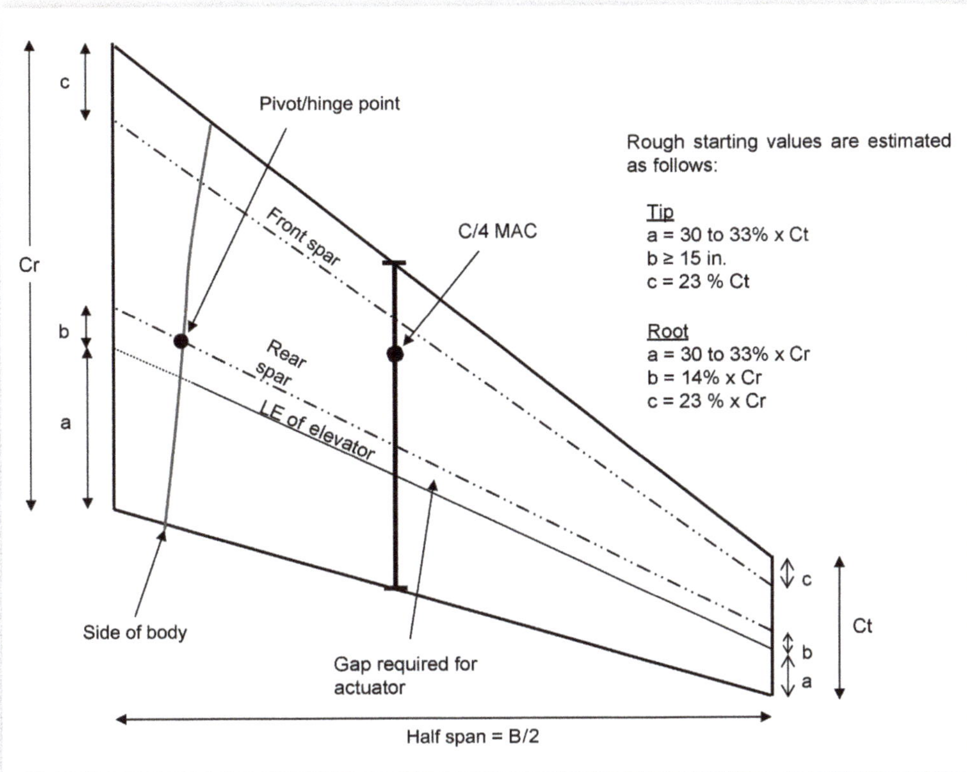

Michael Derrer Fuchs/Shuttersto

8

Vertical Tail Layout

The Vertical Tail Volume Coefficient

The vertical tail helps provide the jet transport with lateral stability. Much like the horizontal tail, the shortest member of the family usually sets the area of the vertical tail (VT). In the case of most conventional airplanes with wing-mounted engines (B777, A330, and MD-11), the vertical tail and rudder are sized by the one-engine out-control requirement. In the case of aft fuselage-mounted aircraft (CRJ-700 and G-V), the vertical tail is usually sized by lateral stability requirements.

The method we will use to calculate the vertical tail area of our aircraft assumes that the vertical tail is sized by the requirement to maintain directional control in the air and on the ground at some minimum control speed. There are several conditions that will need to be checked for both stability and control. Smaller airplanes that are required to have good short-field performance can be sized by the minimum control speed on the ground, while larger transports are typically sized by minimum control speeds in the air. Aircraft with aft-mounted engines such as the CRJ-700 series and business jets such as the Gulfstream V are usually sized by stability considerations.

Our simplified approach to the problem will get us in the ballpark for a vertical tail area. An alternate method to guesstimate the vertical tail area for such aircraft (with aft fuselage-mounted engines) will be presented at the end of this chapter.

We can approximate the minimum control speed from the following chart in Figure 8.1.

FIGURE 8.1 Minimum control speed.

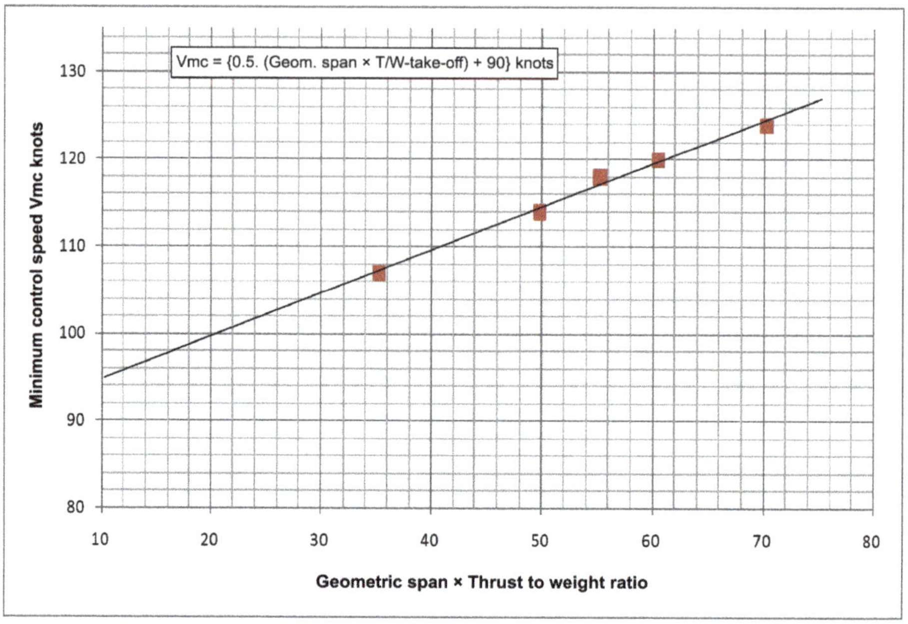

The vertical tail geometric aspect ratios for jets are on the order of the following:

- AR = 1.5 to 1.8 wing-mounted engines (low horizontal tail and T-tail) (B777, A330, B747, MD-11, Tu-334, Do-328 Jet, etc.)
- AR = 1.0 fuselage-mounted engines with T-tails (B727, CRJ-700, Gulfstream, etc.)

The above values for the aspect ratio are concerned with the geometric aspect ratio. Much like a wing with a wingtip device, any sort of endplate effect such as that provided by a horizontal tail at the end of the vertical tail helps the vertical tail increase its effective aspect ratio. The fuselage area just below the vertical tail also contributes to lateral stability and has the effect of increasing the effective VT aspect ratio. This effective aspect ratio can be estimated as follows.

For a conventional empennage arrangement with a low-mounted horizontal tail

$$\text{AR effective} = \text{AR} \times 1.5 \tag{8.1}$$

For a T-tail empennage arrangement with a high-mounted horizontal tail

$$\text{AR effective} = \text{AR} \times 1.8 \tag{8.2}$$

The effective aspect ratio is more pronounced for T-tail configurations because the vertical tail in effect has the horizontal tail at the top and the fuselage below, both of which contribute to the endplate effect.

The figures below illustrate the different forces acting on the airplane with the critical engine failed. In a twin-engine jet airplane, the critical engine can obviously be either one; however, in the case of the four-engine aircraft the critical engine will usually be the outboard most engine. As far as tri-jets such as the MD-11 are concerned, the most critical engine will be either of the wing-mounted engines.

For simplicity, we will ignore the drag of the windmilling engine. We will also assume that the thrust will be 90% of the maximum static thrust due to the engine's thrust lapse rate with increasing Mach number.

Essentially, the lateral yawing moment produced by the vertical tail will have to balance the opposing yawing moment of the remaining engine(s).

Because we have not drawn the aircraft geometry, we can make some assumptions on the location of the engines in terms of the percent of the wing semi-span. The following figures can be used to get approximate figures of engine location in order to get the process started.

FIGURE 8.2a Vertical tail sizing for one engine inoperative—twin jets and quad jets with wing-mounted engines.

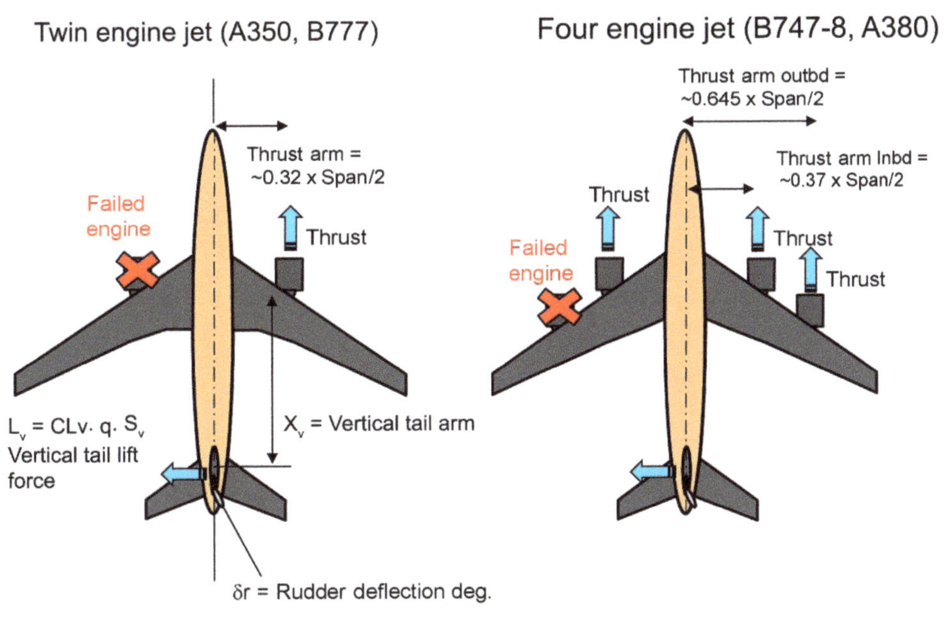

S_v = [Thrust × Thrust arm] ÷ [CLv × Q × X_v]
q = Dynamic pressure at Vmca
CLv = Lift Coefficient of vertical tail
X_v = Distance from quarter chord MAC of vertical tail to quarter chord MAC of the wing

FIGURE 8.2b Vertical tail sizing for one engine inoperative—tri-jets and twin jets with tail-mounted engines.

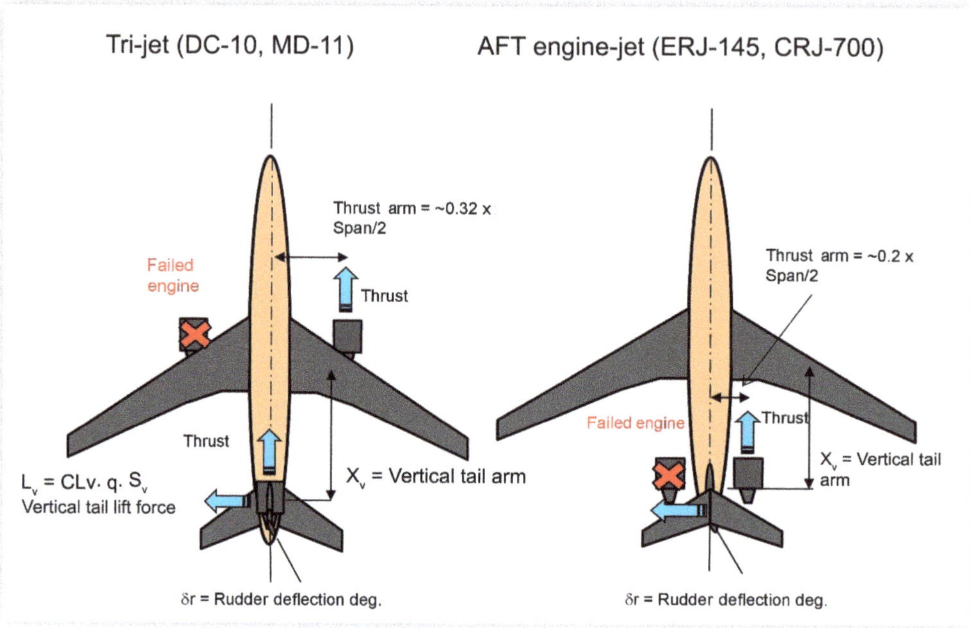

The vertical tail area is estimated as follows:

$$S_v = \left[\text{Thrust} \times \text{Thrust arm}\right] \div \left[C_{Lv} \times Q \times X_v\right] \tag{8.3}$$

where

Q is the dynamic pressure at Vmc = 0.5 × r × Vmc^2. Here, we will assume sea-level density.

X_v is the Vertical tail moment arm (from quarter-chord MAC of wing to quarter-chord MAC of the VT)

= 0.395 × L_f wing-mounted engine configuration with T-tail

= 0.435 × L_f wing-mounted engine configuration with low tail

= 0.385 × L_f aft engine configuration with T-tail

= 0.410 × L_f aft engine configuration + low conventional tail

Thrust arm = 0.32 × wing semi-span without wingtip device (twin-engine airplanes such as B777, A330, and ERJ-170)

= 0.32 × wing semi-span without wingtip device (tri-engine airplanes such as MD-11)

= 0.345 × wing semi-span without wingtip device (inboard engine of four-engine airplanes, e.g., A380)

= 0.63 × wing semi-span without wingtip device (outboard engine of four-engine airplanes, e.g., A380)

= 0.20 × wing semi-span without the wingtip device (aft fuselage-mounted engines such as ERJ-145)

Note: In the case of the four-engine airplane, the outboard engine position would be the most critical engine in terms of the maximum yawing moment generated.

Thrust arm = Thrust moment arm from the engine to fuselage centerline

C_{Lv} is the lift coefficient of the vertical tail

To estimate the lift coefficient of the vertical tail, we will first need to estimate its lift curve slope.

The lift curve slope can be easily approximated using the following equation:

$$\text{CLa} - \text{vt} = \left[a_0 \times \text{AR effective}\right] \div \left[\text{AR effective} + \left\{2 \times \left(\text{AR effective} + 4\right) \div \left(2 + \text{AR effective}\right)\right\}\right] \quad (8.4)$$

where $a_0 = 0.110$ is the approximate airfoil section lift curve slope.

The rudder is the movable surface on the vertical tail that provides lateral control. A brief evaluation of the modern transonic transport shows that the rudder usually has a constant chord ratio that ranges between 30 and 33% of the total chord. The higher chord ratios (those in excess of 33%) usually provide for a more powerful control surface, but this has to be traded against larger actuators to cope with the higher hinge moments and also the higher loads on the vertical tail attachment joints. Deflecting the rudder has the same effect on the vertical tail as a plain flap has on a wing.

Most modern transports have single-hinged rudders, which may or may not have an additional tab. In some cases, however, the rudder can be double-hinged, which can allow for a higher maximum deflection.

For initial sizing purposes, a rudder deflection of about 27° is a good estimate. A rudder with a tab is more effective, and a 33°-deflection is a good first estimate for this sort of configuration. In the case of a double-hinged rudder, a deflection of 36° is probably a good initial number. Again, in early product development it is probably a good idea to limit rudder deflection to less than 30° and allow the vertical tail area to size up. Double-hinged rudders and tabs, etc., add complexity and are usually fixes to compensate for unplanned thrust increases later in the program.

The FARs typically allow a 5°-bank into the good engine. However, for simplicity at this early stage of design no bank angle credit is assumed because we do not have a clear idea about whether the minimum control speed in the air or on the ground is more limiting.

The lift coefficient of the vertical tail can now be approximated as follows:

$$\text{CLvt} = \text{CLa-vt} \times \left(0.6 \times \delta\text{r}\right) \quad (8.5)$$

$$Q = \text{Dynamic pressure} = 0.5 \times \text{density at sea level} \times \text{Vmc}^2 \text{ in lb / ft}^2$$

$$\text{Thrust moment} = \text{Thrust} \times \text{Thrust arm}$$

Even though the initial family member will have a maximum thrust that is lower than the stretch member, it is good practice to protect the capability of the vertical tail for the maximum potential thrust available. This is done to protect for any future growth in gross weight for the initial model. This is common practice as inevitably there will be customers who want to expand the performance envelope of the initial aircraft in an aircraft family. Furthermore, in the initial assessment no credit for thrust reduction due to speed is taken.

Hence, we now have all the parameters to estimate our vertical tail area.

$$S_v = \left[\text{Thrust} \times \text{Thrust arm}\right] \div \left[\text{CLv} \times Q \times X_v\right] \tag{8.6}$$

The volume coefficient is calculated as follows:

$$V_v = (X_v \times S_v)/(S_{ref} \times b_w)$$

where

X_v is the distance of the VT quarter-chord MAC to the center of gravity of the aircraft. For practical purposes in the early design phase, the quarter-chord of the wing MAC is used instead of the center of gravity.

S_v is the vertical tail area

S_{ref} is the wing area

b_w is the effective aerodynamic wingspan

Reference [29] has the vertical tail areas of current transports. It also serves as a good check for a particular aircraft type. The tail volume coefficients are calculated using the same approximations of tail arms mentioned previously, since accurate tail arm data are proprietary and not readily available (Table 8.1).

TABLE 8.1 Vertical tail characteristics of existing jet transports.

Aircraft type	Wing area S_{ref} (ft^2)	VT area S_v (ft^2)	Rudder type	Rudder span as fraction of VT span	Effective aerodynamic wingspan (ft)	VT volume coefficient (estimated)
A320-200	1320	231.4	Single hinge	1.0 (full span)	115.3	0.0806
Fokker 70	1006	132.4	Single hinge	1.0	92.13	0.0497
A340-500	4704	513	Single hinge	1.0	209.0	0.0483
A330-200	3892	513	Single hinge	1.0	198.0	0.0543
B777-200ER	4605	573	Single hinge with tab	1.0	199.92	0.055
MD-11	3648	605	Double hinge	~0.57	176.0	0.0689

Sizing the Vertical Tail for Aircraft with Aft Fuselage-Mounted Engines

In the case of aircraft with aft fuselage-mounted engines, it is usually the lateral stability criteria that size the vertical tail and not so much the engine out-control requirement. Here, we can use an empirical formulation that utilizes some fuselage and wing parameters.

Much like the horizontal tail sizing exercise, we can define a K-factor,

$$K_f = \left[H_f^{1.05} \times L_f\right] / \left[S_{ref} \times \text{Wingspan}\right] \tag{8.7}$$

where
L_f is the fuselage length (ft)
H_f is the fuselage height (ft)

At this stage of the sizing process, we know a fair amount about the configuration, and what we do not know, we can assume. Typically, if we have not drawn the cross section by this stage, we can assume that the diameter of the fuselage is equal to the height. Otherwise, we should use the height of the fuselage if we know it.

$$X_v = 0.385 \times L_f \text{ aft engine configuration with T-tail}$$

Note: The wingspan should be the effective aerodynamic wingspan.

$$V_v = 11.55 \times K_f - 0.0825, \text{where } K_f = \left[Hf^{1.05} \times L_f\right] / \left[S_{ref} \times \text{Wingspan}\right] \tag{8.8}$$

However, the volume coefficient can also be written as follows:

$$V_v = \left[X_v \times S_v\right] / \left[S_{ref} \times \text{Wingspan}\right] \tag{8.9}$$

By equating the two equations for the VT volume coefficient, we can solve for the VT area.

$$S_v = 11.55 \times \left[H_f^{1.05} \times L_f\right] / X_v - 0.0825 \times \left[S_{ref} \times \text{Wingspan}\right] / X_v \tag{8.10}$$

Aspect Ratio

VT aspect ratios range from 1.0 to 1.8. The lower number is usually associated with T-tails. The rationale being that a lower aspect ratio results in a more structurally efficient design which is rigid and can accommodate the mass and loads of the horizontal tail.

The higher aspect ratio VT found on conventional empennage arrangements makes for a more aerodynamically effective, but heavier structure. Aerodynamic efficiency is realized via an increase in the lift curve slope or CL_{α}. Hence, for a given angle of attack the VT will be more powerful. This helps with reducing cruise profile drag and increases control power. This decrease in area is typically balanced against increased structural weight for the more slender surface. Typically, the aspect ratio is limited to 1.8 or thereabout.

Taper Ratio

VT taper ratios range from 0.25 to 0.8. The higher number is usually associated with T-tails. The reason is that the high taper ratio allows for a larger tip chord, which can accommodate the HT. The higher taper ratio can also make for a more efficient but potentially heavier empennage. In the context of the vertical tail layout, the taper ratio is often used to help integrate the different components such as the elevator actuation with adequate box chord to manage the structural loads.

Thickness-to-Chord Ratio and Quarter-Chord Sweep

Thickness-to-chord ratios typically range from 9 to 10%. The quarter-chord sweep of the vertical tail for conventional designs is typically set at 40°, which is typical of designs cruising at 0.75 Mach or greater. The aim here is to make sure that the combination of sweep and thickness allows the VT to maintain its control effectiveness long after the wing is seeing shock-induced separated flow. In the case of T-tail aircraft, an increased sweep on the order of 45 to 50° may be necessary to provide added VT box thickness to react to the added HT loads. The increased sweep also increases the tail arm of both the VT and HT, particularly for aft engine configurations where the wing is further aft on the fuselage. In other words, the sweep and thickness of the entire empennage (VT and HT) will have to be played with to get a net weight and drag benefit while meeting the stringent stability and control requirements.

Ice Protection

One of the sizing criteria that can impact the vertical tail performance is whether it does or does not have an ice protection system or if the geometry (sweep and thickness) of the vertical tail is more conducive to ice formation. This can affect both weight and drag if it becomes necessary to size the VT up in area to maintain stability and control without the added system complexity of an anti-ice system.

FIGURE 8.3a Vertical tail layout.

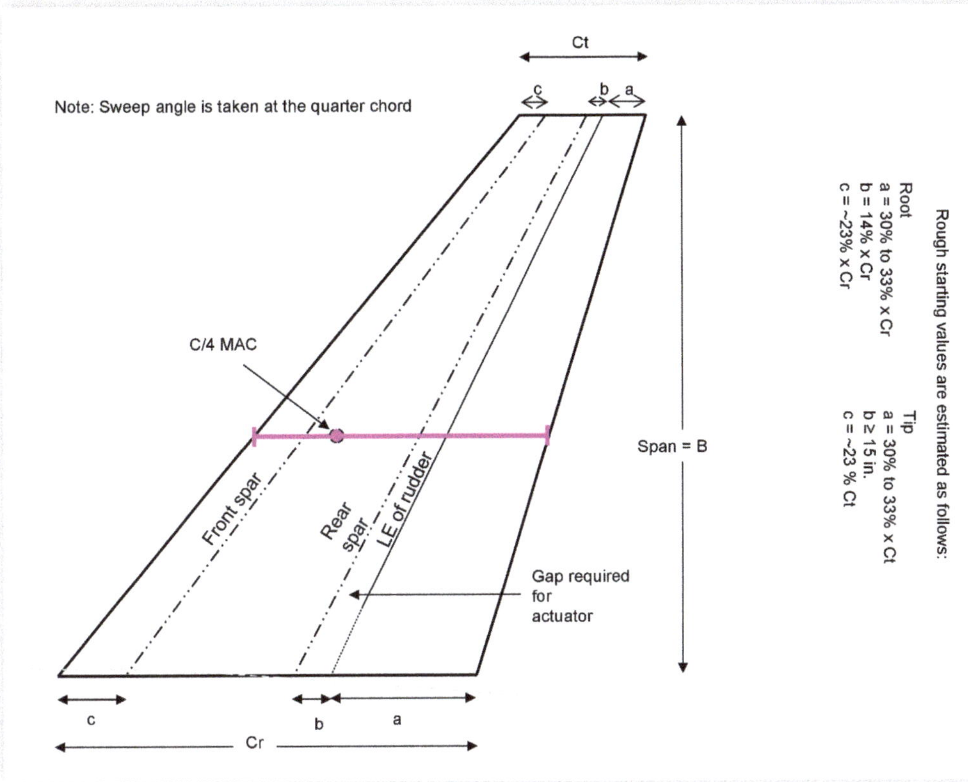

© Zarir D. Pastakia.

FIGURE 8.3b Vertical tail layout (continued).

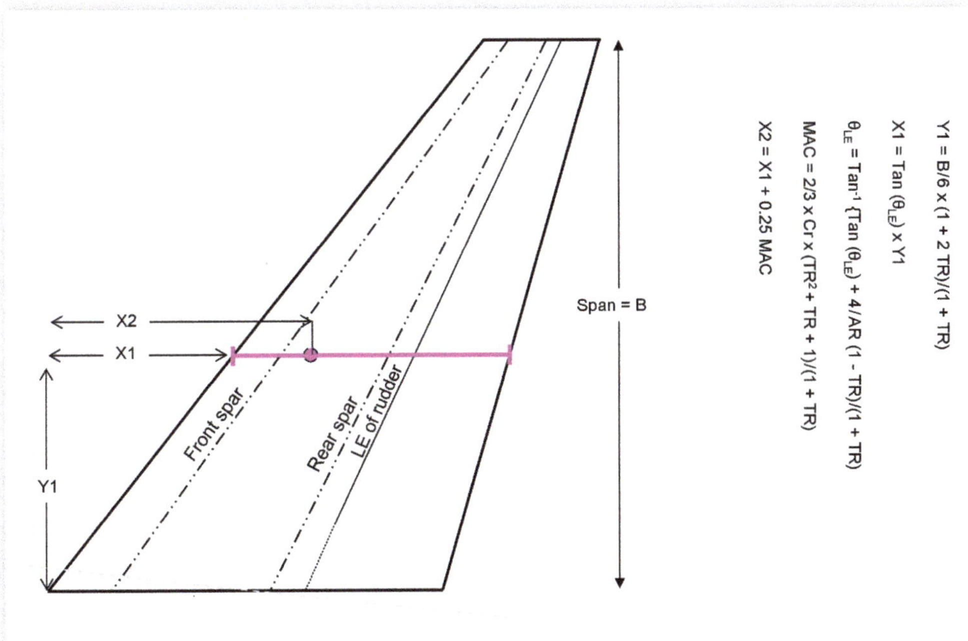

© Zarir D. Pastakia.

FIGURE 8.4 Vertical tail of the 777-9.

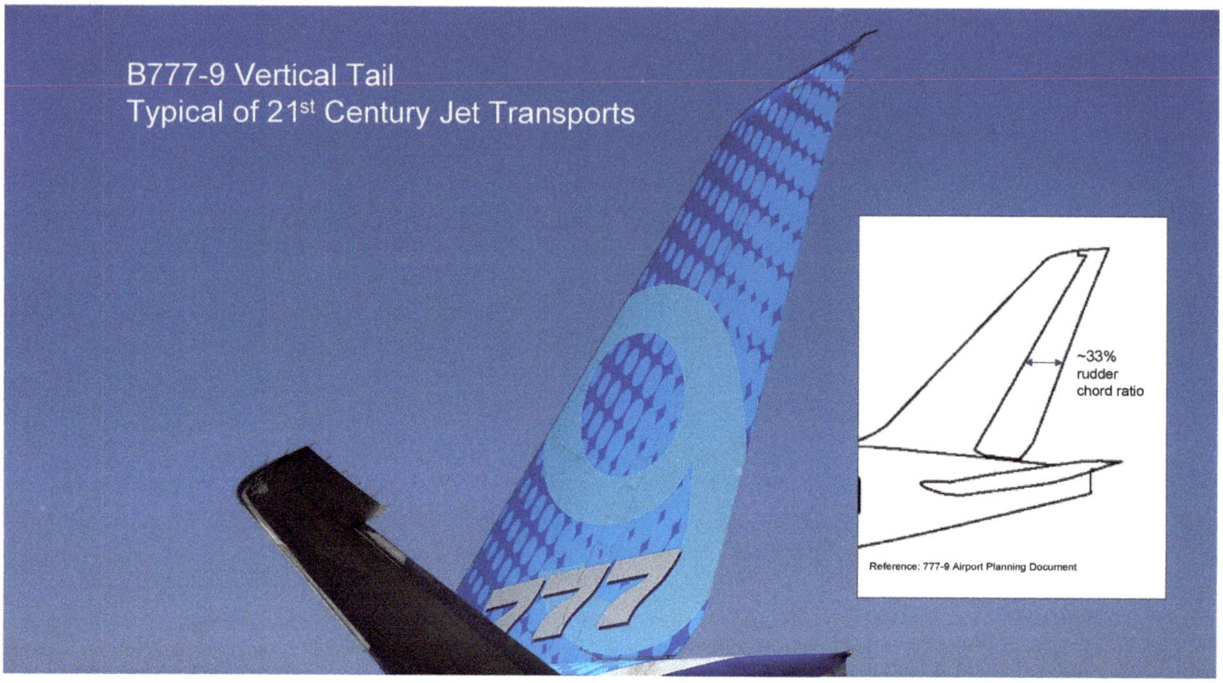

Uwe Deffner/Alamy Stock Photo.

Media_works/Shutterstock

9

Propulsion and the Jet Engine

The Jet Engine

While this is an airplane design book, it goes without saying that some of the largest gains in airplane design have come via the propulsion system. The jet engine embodies the best that all the aerospace engineering discipline has to offer in terms of aerodynamics, materials, structures, and system technologies. The core inside a modern jet engine sees temperatures approaching 2000°C, while the engine itself is operating in an environment where the temperature can fluctuate between 35°C on the ground and –55°C at altitude. Modern control systems enable automatic control and resolution of minor problems autonomously. Because the subject of the book is jet transports, the focus of this chapter will be on the jet engine and its design features.

Simplistically, the jet engine works on the following principle:

- Suck—Where the air is taken into the engine.
- Squeeze—Air is compressed to raise the pressure and temperature.
- Bang—Air is combusted, and chemical energy is converted into thermal energy.
- Blow—Air is exhausted via a nozzle at a higher velocity.

The focus of early jet engine design was to produce thrust for military aircraft. The military mission demanded a tightly packaged propulsion unit that produced high thrust for speed and maneuverability. Thrust is produced by taking in a mass of air, accelerating it, and exhausting it at a higher velocity. Propulsive efficiency increases as the flight velocity and jet exhaust velocity converge. Hence, for fast supersonic military jets propulsive efficiency was achieved via a higher flight velocity.

For early transonic jets such as the DE Haviland Comet, which cruised in the Mach 0.78 cruise Mach number range, propulsive efficiency decreased due to the flight speed being a lot slower than the jet exhaust velocity. In the 1950s, early jet transports were still experimenting with engine position and the compact jet engine allowed flexibility in this regard, especially with early British designs, which sought to bury the engine within the wing roots, allowing for a compact package.

Net thrust or FN can be defined as follows:

$$\text{FN} = \{\text{Mass flow rate of air} + \text{Mass flow rate of fuel}\} \times \text{Jet exhaust velocity} - \{\text{Mass flow rate of air} \times \text{Velocity of flight}\} \quad (9.1)$$

$$\text{Mass flow} = \text{Density} \times \text{Area} \times \text{Velocity} \quad (9.2)$$

The negative term in the equation is also called ram drag and has to do with the aerodynamic drag of the engine itself as it pushes the airplane through the air. When the aircraft is not moving, this term goes to zero, and what remains is the gross thrust of the engine.

Hence, early jet engine design focused on creating thrust by pushing a small amount of air (in a compact engine) by a large amount of speed (exhaust velocity). However, as fuel efficiency and range grew in importance, the high bypass ratio turbofan engine was invented to improve propulsive efficiency. The bypass ratio is the ratio of the amount of air that bypasses the core of the engine versus the amount that is processed through the core.

The turbofan enabled efficiencies by moving a larger mass of air (via a large fan) at a slower velocity. This sort of engine design enables improved propulsive efficiency by slowing down the jet exhaust velocity. Hence, as material science and structural design improved, a push was made to larger and larger fan diameters. However, there are penalties to be had with increasing fan diameters—the obvious one being weight. Also, as fan size increases, the fan tip speeds can approach the speed of sound, causing aerodynamic losses. Through the years, improvements in fan blade design (aerodynamic and structural design) have helped manage aerodynamic losses at the tips. As engine core design has improved via improved materials, better packaging, and better cooling, increases in bypass ratio have also been possible via shrinking the core while holding the fan diameter the same.

The latest improvement in pushing for higher propulsive efficiency is the Pratt & Whitney Geared Turbofan (GTF). This design allows for the integration of a gear mechanism within the engine that allows the fan to rotate at a slower speed, thereby allowing an even larger fan diameter for a given thrust requirement.

As engine fan diameters increase, so does the weight of the nacelle or cowl that surrounds the engine. The nacelle provides a smooth aerodynamic body to wrap the engine and other associated accessories. The part of the engine that provides the ability to reverse the airflow on the ground (commonly referred to as a thrust reverser) is also driven by the size of the cowl.

The thrust reverser and fan cowl can weigh almost as much as the engine itself so the weight penalties associated with a larger fan can be significant. The GTF engine combats some of this weight increase by allowing for a reduced-length engine.

In the end, the better-performing engine is the one that brings efficiency via a balance of aerodynamics, weight, and thrust-specific fuel consumption.

Figures of Merit for Engine Choice

- Engine investment and direct operating cost.
- Fuel burn (combination of weight, drag, and specific fuel consumption).
- Noise.
- Emissions (carbon, nitrogen oxide, etc.).
- Availability—new engine development programs can take up to seven years from early requirements to having an engine ready for flight test.
- Reliability and maintenance cost.

The Inlet

While the inlet is not part of the engine itself, it is an integral part of the propulsion system and has a significant bearing on the performance of the engine and the aircraft as a system. The main task of the inlet is to decelerate the air from flight speed to an acceptable velocity (usually Mach 0.5) before it meets the fan and the compressor. Decelerating or diffusing the air also helps increase the pressure of the air, thereby helping the overall engine's efficiency. This is because less work must now be done by the compressor in raising the pressure further. Inlets are designed to be efficient at the design point, such as cruise, but to also minimize drag during an engine failure on take-off.

The Fan and Compressor

The fan is a large, low-pressure compressor found at the front of a civil turbofan engine and works on the total airflow through the engine. The fan flow on a modern turbofan engine produces about three-fourths or 75% of the engine's thrust.

The splitter is a feature that as described splits the incoming airflow into the bypass stream and core stream. The low-pressure and high-pressure compressor stages are found just aft of the fan. They compress the air that enters the core of the engine. The main task of the compressor is to raise the pressure of the air passing through it before it enters the combustor. All modern civil turbofan engines have an axial compressor system due to the compression efficiency of this arrangement. Each stage makes up a rotating airfoil system that progressively compresses the air as it moves toward the combustion chamber. Compressing the air also raises the temperature of the air molecules in preparation for the combustive process.

The low-pressure (LP) compressor stage is upstream, downstream of the fan, and it is driven by the low-pressure turbine. The high-pressure (HP) compressor stage is downstream of the low-pressure stage and is the final stage before the air is sent into the combustor. It is driven by a high-pressure turbine.

Some air is bled off from the compressor to pressurize the cabin and provide warm air for anti-icing the leading edge of the wing and nacelle intake.

The compressor pressure ratio is the ratio of the total pressure at the compressor exit and compressor entry.

$$\text{Total pressure} = \text{Pressure at compressor exit} \div \text{Pressure at compressor entry}$$

The compressor pressure ratio has a direct bearing on thrust and overall engine fuel consumption and weight efficiency. This is because increasing the pressure ratio typically requires an increased number of stages and increased stress levels due to the increased pressure and temperature, all of which necessitate an increase in engine weight.

Compressor design is a multidisciplinary design problem involving optimization of aerodynamics, material systems, manufacturing methods, mechanical design, and cost.

The overall pressure ratio (OPR) is a function of the compressor pressure ratio. It is defined as the pressure between the fan inlet stage and the last compressor exit stage. It is often used as a measure of the technology within an engine. OPRs have been increasing steadily with time, as material science and component cooling architectures have improved.

FIGURE 9.1 Overall pressure ratios with time.

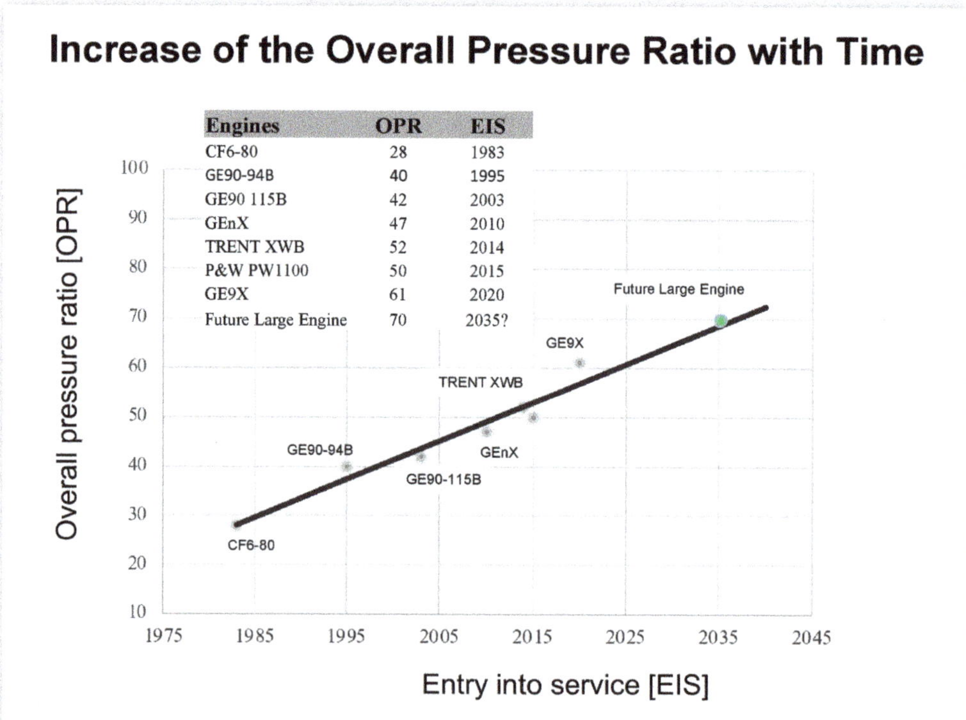

The Combustion Chamber

As the air leaves the compressor, it is traveling at almost 300 knots (345 mph). This makes combustor design extremely challenging as aviation fuel has to be mixed in with the air and burned—a problem akin to lighting a match in a hurricane. It is in the combustor where chemical energy (aviation fuel) is converted into thermal energy.

Heat energy must be released into the turbine in a uniform way. Some of the key requirements that must be met by the combustor are as follows:

- High combustive efficiency and low emissions
- Ensuring the flame stays alight over a wide environmental range of rain, hail, and off-design conditions
- Ability to restart the engine at high altitude
- Ensure a durable design that is both low in weight and cost

As material systems, engine cooling systems, and overall design have continued to improve, turbine entry temperatures (TETs) have continued to increase.

The Turbine

The first task of the turbine system is survival, Glowing red-hot, the blades operate in temperatures well above their melting point... The turbines second task is to drive the compressor. (Reference: Rolls Royce, The Jet Engine).

A conventional turbine system is made up of high-pressure turbine stages just downstream of the combustor and a low-pressure turbine system, which can comprise various stages. The turbine system is made up of rotating airfoil blades and static vanes. The entire turbine system is contained in a divergent casing.

The low-pressure turbine provides power to the low-pressure compressor and the fan, while the high-pressure turbine drives the high-pressure compressor that compresses the core air before it enters the combustor. A small portion of the power from the high-pressure turbine is used to drive fuel pumps and provide electricity and hydraulic power for the aircraft.

Some of the key requirements that must be met by the turbine are as follows:

- Providing the required thrust
- Minimizing cost
- Minimizing weight
- Minimizing emissions
- Minimizing fuel consumption

The turbine design is also a multidisciplinary design problem involving the optimization of aerodynamics, material systems, manufacturing methods, mechanical design, and cost.

The Nozzle

Once the air passes through the turbine, it is delivered back to the atmosphere through the nozzle. Most nozzles on modern turbofan engines are convergent nozzles, where the air exiting the turbine is accelerated to provide thrust. Since a modern high bypass ratio engine has a bypass stream (fan flow) and a core stream, these engines have a primary exhaust nozzle (core airflow that exits at hot temperatures and higher velocities) and a secondary fan nozzle (where the exhausted air is slower moving and at colder temperatures).

Nozzle design on a modern engine is driven by the following:

- Low community noise and noise in the cabin
- Efficient thrust recovery
- Weight
- Aerodynamic drag

The cone protruding from most modern engine designs is there to allow further expansion of the jet, causing a pressure profile that has the effect of generating additional thrust forces on the cone.

FIGURE 9.2 Typical direct drive turbofan definition and dimensions.

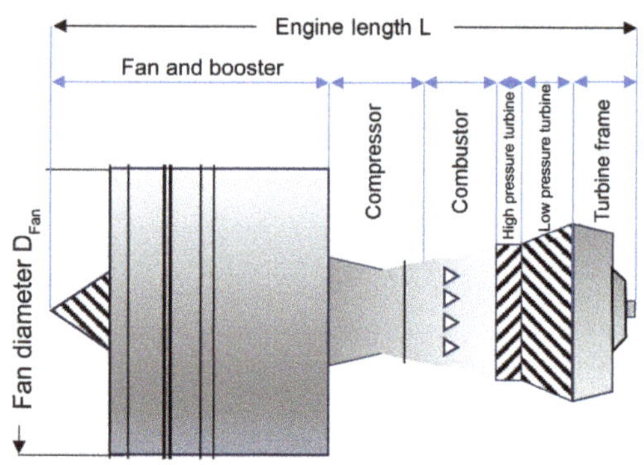

Engine	Type	Fan diameter D_{Fan}	Engine length L	Length/ fan diameter
CFM Leap-1A	DD	78	135.6 est.	1.74
PW 1100	GTF	81	133.9	1.65
GEnX	DD	111	194.9	1.76
Trent XWB	DD	118	210.0	1.78

Reference: Wikipedia, Type Certificate Data Sheets & Delta Website
Note: All dimensions are in inches
GTF = Geared turbo fan
DD = Direct drive
Estimating Engine Dimensions
Fan diameter $D_{Fan} = 0.226 \times [Fn^{0.5} \times BPR^{0.25}] + 4.71$
Engine Length $L = 1.76 \times D_{Fan} \times k$
k = 1 for a direct drive engine
k = 0.9375 for a geared turbo fan engine

Fn = Thrust = 0.977 x Sea Level Static Thrust
BPR = Bypass Ratio

FIGURE 9.3 Engine nacelle definition and dimensions.

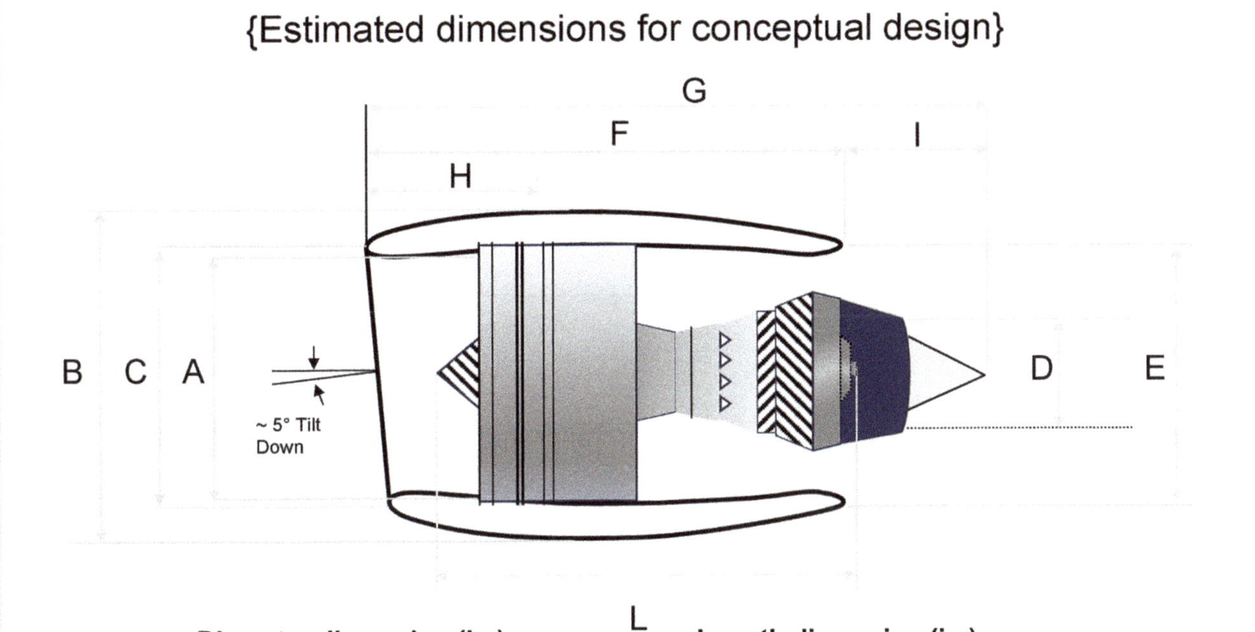

Diameter dimension (in.)

A = Fan diameter = 0.226 × $[Fn^{0.5} \times BPR^{0.25}] + 4.71$
B = Nacelle diameter = A × [2 – 0.06 × Ln (Fn)]
C = Inlet dia. = 0.8 x B
D = Nozzle core diameter = 0.4 x B
E = Fan exit diameter = 0.725 x B

Length dimension (in.)

L = Engine length = 1.76 x D_{Fan} x k
k =1 for a direct drive engine
k = 0.935 for a geared turbo fan
F = Nacelle main cowl length = 1.15 × L
G = Total length = 1.415 × F
H = Inlet length = 0.35 × F
I = Primary nozzle + Plug length = 0.4 × L
J = Plug length = 0.24 × G

Engine Installation

Early jet transports experimented with engine installation schemes that were enabled by the turbojet. Some of these installations are shown as follows.

FIGURE 9.4a Embedded wing installation in a De Haviland Comet.

© Zarir D. Pastakia.

FIGURE 9.4b Aft fuselage-mounted installations of 1960's vintage.

© Zarir D. Pastakia.

Wing-Mounted Location

FIGURE 9.5 Wing-mounted engine installation general trends.

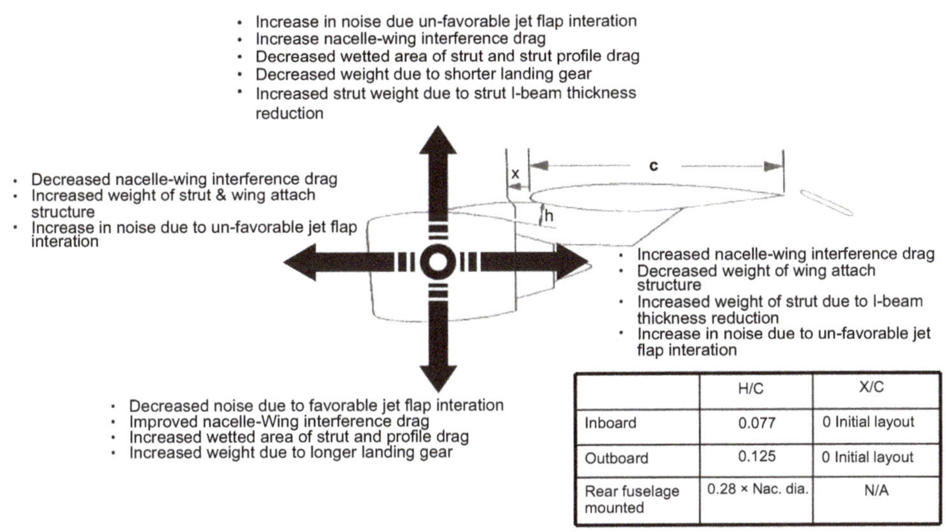

	H/C	X/C
Inboard	0.077	0 Initial layout
Outboard	0.125	0 Initial layout
Rear fuselage mounted	0.28 × Nac. dia.	N/A

- Propulsion units are heavy and have the benefit of providing wing-bending relief.
- The propulsion unit mass is close to aircraft CG longitudinally, allowing for a more balanced aircraft family (stretch and shrink).
- Engine rotor burst is concentrated away from flight controls.
- On low-wing airplanes, in the landing, go around phase, with the flaps retracting, and with the airplane trimmed to be nose up, increasing thrust has a destabilizing (pitch-up) moment as the thrust line is below the center of gravity. The high-wing airplane alleviates this destabilizing pitch-up tendency.
- On a low-wing airplane, because the engine is closer to the ground it is more susceptible to foreign object damage (FOD) and hence not well suitable for gravel runways. Figure 9.6 shows the ground clearance of engines on some of the modern transports of today.

FIGURE 9.6 Engine ground clearance.

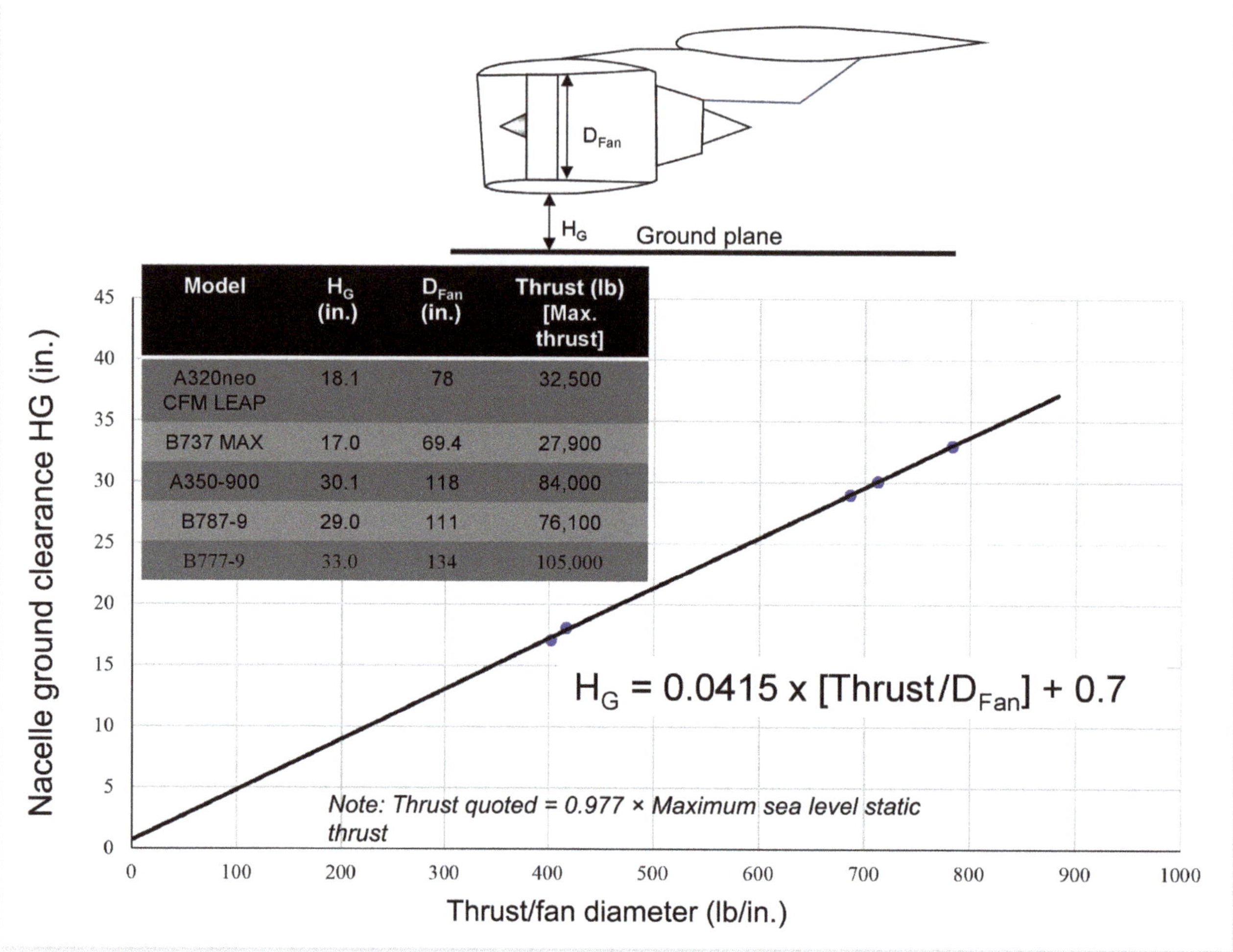

Model	H_G (in.)	D_{Fan} (in.)	Thrust (lb) [Max. thrust]
A320neo CFM LEAP	18.1	78	32,500
B737 MAX	17.0	69.4	27,900
A350-900	30.1	118	84,000
B787-9	29.0	111	76,100
B777-9	33.0	134	105,000

FIGURE 9.7 FAR AC150 runway sign clearance and location.[1]

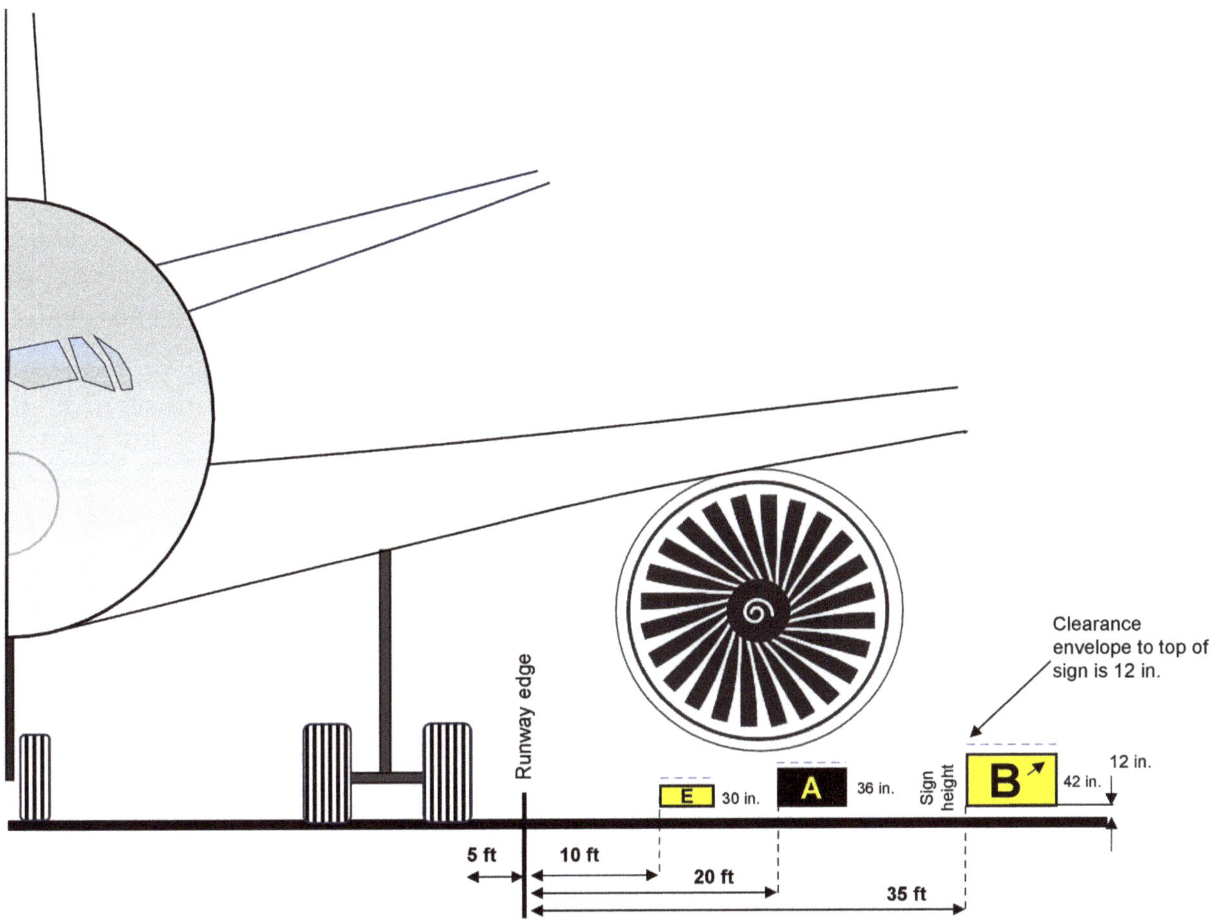

Created by Zarir D. Pastakia_MicroOne/Shutterstock.com.

[1] https://www.faa.gov/documentlibrary/media/advisory_circular/150_5340_18f.pdf

- Engine ground clearance relative to airport taxiway signs is also an important consideration and is governed by airworthiness circular AC150.
- On a low-wing airplane, high bypass ratio engines with large fan diameters can drive a longer landing gear, which comes at the expense of increased structure and system weight.
- A longer landing gear also requires a longer trunnion or pintle, which has a direct effect on wing design, requiring more wing area or less wingspan to meet integration requirements, specifically integration of the main landing gear attach structure (gear beam) and the inboard spoilers and flaps.
- Wing gulling may be required, and this can add wetted area (more drag) and be challenging for wing build, requiring thick wing skins to be formed to complex curvatures—not an insignificant challenge.
- Wing-mounted engines can impact shorter-body family members, which have the added challenge of integrating forward passenger and cargo doors.

- Other integration challenges include placing the air data system (airspeed pitot tubes and static ports) along the forward fuselage. This system requires clean air for accurate measurements and must not be influenced by the aerodynamics of the engine nacelle and/or engine reverse thrust, which can cause this system to be contaminated.
- On a low-wing airplane, a short fuselage with wing-mounted engines is also critical for nose gear collapse, which has a requirement to not damage the fan. This could limit how short the fuselage can be and/or require the engine to be tucked up higher than the wing, thereby compromising wing aerodynamics.
- Wing-mounted engines on short fuselages help with the sizing of relatively shorter slides on the front door due to the main landing gear being more centrally located versus an aft engine-mounted configuration.
- Cabin noise increases aft of the wing requiring additional noise shielding on the engine and/or the fuselage necessitating an increase in weight. This noise effect is due to the high-velocity air coming out of the exhaust of the engine.

Aft Fuselage-Mounted Location

- Well suited for smaller regional jets and business jets.
- Allows for a shorter landing gear that is lighter and places the aircraft closer to the ground.
- The short main landing gear drives a smaller trunnion or pintle, which makes for easier integration of the main landing gear attach structure with the inboard spoilers and flaps.
- Improved inboard wing integration provides the opportunity to bias more wing area outboard for improved aerodynamics (assuming a fixed-wing area).
- The propulsion unit mass moves the center of gravity aft, thereby causing the wing to move further aft to balance the aircraft. This could make creating an aircraft family more challenging.
- Stretching the body can become challenging given the large shift in the center of gravity between an empty aircraft (CG aft) and a loaded aircraft (CG moves forward).
- Shrinking the body can be challenging to make sure there is enough room for a door and escape slides between the wing- and aft-mounted engines.
- Engine rotor burst is concentrated around flight controls running to the empennage.
- The engine is mounted further off the ground and is less susceptible to foreign object damage (FOD) and hence more suitable for gravel runways.
- Large fan diameters have a minimal impact on the wing in terms of wing "gulling."
- There is less of an integration challenge in placing the air data system (airspeed pitot tubes and static ports) that requires clean air for accurate measurements.
- Fuselage-mounted engines drive the sizing of relatively longer slides on the front door due to the gear being located further aft relative to a wing-mounted configuration. A longer slide that meets an adverse attitude (nose high up when tipped back) is then shallower when deployed during a normal attitude evacuation (too flat to slide down). These design details must be looked at earlier in the design.

Thiago B Trevisan/Shutterstock

10

Pulling It All Together

This is the fun part, where we pull the airplane together. The method described here is by no means gospel. Furthermore, it is an iterative method, and the success of the outcome will depend on the designer's creativity and problem-solving ability. In airplane design, there is no "one-size-fits-all" methodology. The information presented below is simply guidance on how to start the integration process and some tools to make sure the configuration holds together when it is analyzed.

Locate the Wing in Plan View [73, 74, 78]

1. Starting in the plan view, we first locate the wing along the length of the body. We do this by placing the wing quarter-chord point of the mean aerodynamic chord at a percentage of the body length (nose to tail).
2. The table in the attached figure is an estimate of where the wing should be located for different types of aircraft. The underlying assumption on locating the wing is that the aircraft has relaxed static stability and a fly-by-wire flight control system. This is important because a FBW system typically allows a smaller horizontal and vertical tail, which means less weight in the rear of the aircraft. Smaller and lighter tails mean the wing would need to be further forward to balance the aircraft.
3. The values shown in the table are also synonymous with the initial smaller member of the aircraft family. When you stretch an airplane, a good rule of thumb is to stretch it with a 60 to 40% ratio. In other words, you add more length to the

forward fuselage than the aft. The converse is true when you shrink the fuselage length, removing more in the front end than the aft end of the body.

4. The rationale for the forward-biased stretch is that the aft fuselage tends to be heavier than the forward fuselage, because the structure must react to the loads from the empennage.
5. Single-aisle aircraft tend to have the wing positioned further forward on the fuselage in terms of percentages. This is because on a single-aisle aircraft the net weight of the wing and propulsion system is relatively higher as a percentage of the operating empty weight versus a twin aisle. Single-aisle aircraft also have a higher payload fraction as a percentage of the operating empty weight when compared to the twin aisle. Both these factors tend to bias the wing further forward on a single-aisle aircraft.
6. Rear-engine aircraft tend to push the wing aft to balance the weight of the engines. Besides the Boeing MD-11 and the Lockheed L10-11, most rear-engine aircraft tend to have T-tails. T-tails can also be heavier due to the vertical tail, which has to support the horizontal tail, but they also push the horizontal tail further aft than a conventional low-mounted horizontal tail.

FIGURE 10.1 Locating the wing in plan view.

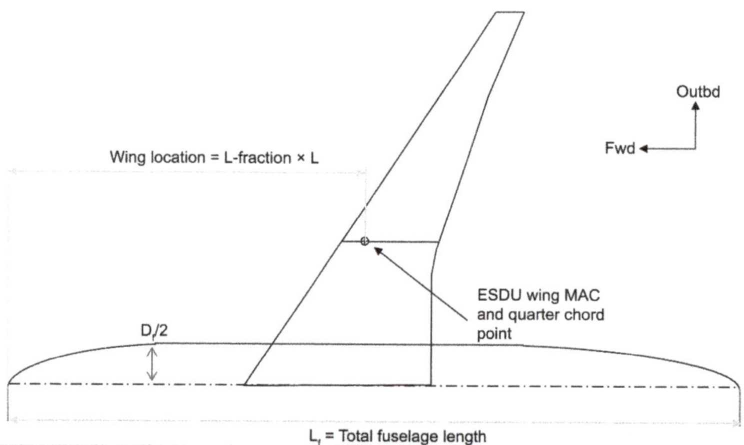

Aircraft configuration type (Assumes first of model and a fly by wire flight control system)	Quarter chord ESDU MAC position as a fraction of fuselage length (L-fraction)
Tri-jet with wing & fuselage mounted engine (MD-11)	47.5%
Twin aisle single deck and double deck aircraft (B777, A380)	47.5%
Twin aisle deck & half aircraft (B747)	47.5%
Single aisle 2 engine aircraft (A320, A220, Embraer E2 jet)	45.0%
Tri-jet with wing & fuselage mounted engine (MD-11)	55.0%
Jet with fuselage mounted engines only (B717, VC-10)	55.0%

7. Once the wing and the main landing gear wheel well are located, we need to draw in a mixed bay (bay used to house system components related to the air conditioning system) forward of the front spar. Mixed bay lengths range from 35 in. on a single-aisle aircraft to 45 in. on a twin-aisle jet.
8. The mixed bay shares a common bulkhead (4 to 5 in. thick) with the forward cargo bay. The length of the forward cargo bay is set by the number and type of pallet or container desired.

   The aim is to maximize pallet capacity in the forward hold and container capacity in the aft hold. It is good design practice to design the aft hold such that passenger bags fit in the aft hold only, while the forward hold is available to carry revenue cargo.

   A check should be made that the container volume in the aft hold can accommodate no less than 6.5 ft^3 per passenger assuming an 85% packing efficiency. As an example, if a 250-seat jet is designed to carry LD3 containers, which have a volume of 159 ft^3, then the target number of containers required for the aft cargo hold would be $\{250 \text{ passengers} \times 6.5 \text{ ft}^3 \text{ per passenger}\}/\{0.85 \times 159 \text{ ft}^3\} = 12$ LD3 containers.

   Almost all modern twin-aisle jets carry LD3 containers in the aft hold. However, some jets carry different containers in the aft cargo hold such as the 747 (LD1 container), the 767 (LD2 container), and the A320 (LD3-45 container). The A320 is unique in that it is the only single-aisle airliner to date that has container-carrying capability. The Chinese COMAC C919 and the Russian MC-21 single-aisle airliners will also have the ability to carry containerized cargo in the cargo hold. All other narrow bodies have bulk cargo capability only, including the new A220. Most narrow-body airliners typically carry bulk cargo in both the forward and aft hold, and a minimum volume of 7 ft^3 per passenger sets the total volume necessary.

   The A320 gives airlines the flexibility to carry both containers and/or bulk cargo. This flexibility comes at a price of both weight and drag due to a taller cross section and the complexity of a cargo handling system in the lower hold. However, the speed and convenience afforded by interlining containers between aircraft models should not be dismissed, especially in the case of an airline that has a mixed fleet of twin-aisle and single-aisle models. In such cases, utilization can be maximized by minimizing connection times at a hub airport. If the minimized turn time on the ground can lead to an extra daily flight, the added revenue would easily offset the fuel burn advantage of a reduced fuselage cross-sectional area. In the end, this boils down to the airline's ability to take advantage of its operational network.
9. The containers and pallets are arranged so that they are separated by 1 in. between the individual units. Once sized, one should do a check to see how flexible the forward cargo compartment is in accepting passenger containers too. This is because different operators may opt to fill either cargo bay with a mix of containers and/or pallets. Flexibility is the key to airline operations, especially when you consider that an aircraft can be in service for a 20-year period or more. Market conditions (passenger load factors, revenue cargo evolution on a given route, airline interiors, etc.) evolve, and a lot can change in a 20-year service life of an aircraft.

10. The forward cargo compartment shares a common bulkhead (3 to 5 in. thick) with the electronics and equipment bay—commonly called the EE Bay. The EE Bay is home to the various avionics boxes that form the nerve center of the aircraft. A good assumption for an estimate of the EE Bay length is 80 in. It is not unusual to have more than one EE Bay such as is the case with the 787 all-electric aircraft. Hence, when pursuing a system architecture on an aircraft one must fully understand the spatial needs necessary to meet the systems volume and separation requirements for certification.
11. The EE Bay shares a common bulkhead with the nose landing gear wheel well. The forward face of the nose wheel well shares a common bulkhead with the forward radome. The nose wheel well should be long enough to accommodate the length of the nose landing gear with the oleo (shock absorber) in the extended position (typical extension is 18 to 20 in.).

FIGURE 10.2 Fuselage below floor integration.

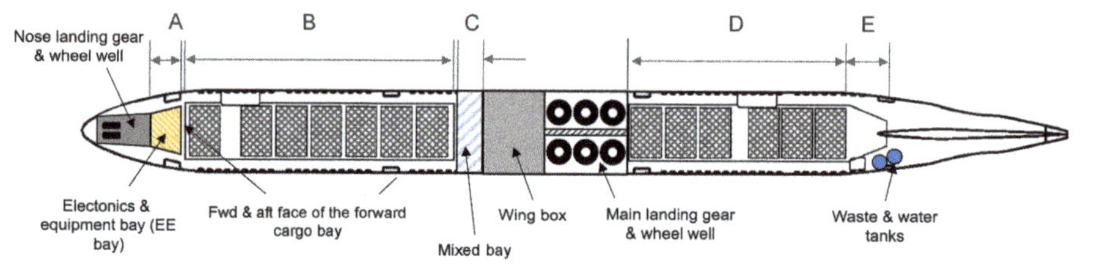

Model	Forward cargo bay length B (in.)	Number of 96 × 125 pallets in forward cargo bay	Aft cargo bay length D (in.)	Number of containers & type in aft cargo bay	Bulk cargo bay length E (in.)	Bulk cargo bay volume (ft^3)
A220-300	277	Bulk only	330	Bulk only	NA	NA
A320 neo	195	Bulk or 3 LD3-45	232	Bulk or 4 LD3-45	154	208
B787-9	618	6	494 est.	16 LD-3	103 est.	402
A350-900	638	6	535	16 LD-3	75	400
B777-300ER	799	8	636	20 LD-3	176	600
A380-800	685	7	705	16 LD-3	110	650

A = Electronics & Equipment Bay. A good initial estimate is 80 in.
C = Mixed Manifold Bay for the Environmental Control System. Typically range from 35 in. for small aircraft to 45 in. for a large aircraft

FIGURE 10.3 Cargo Bay (forward and aft) container arrangement.

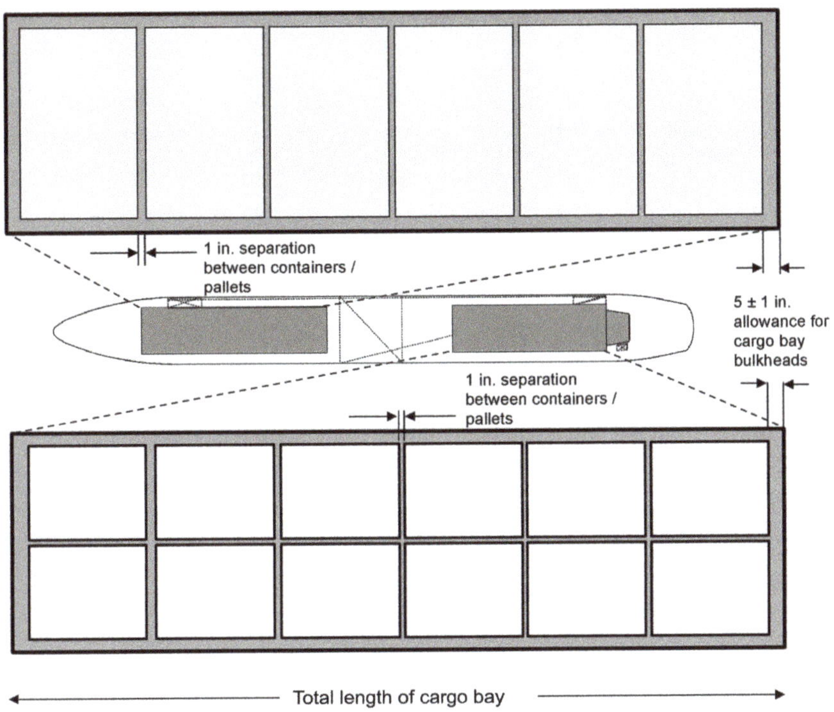

FIGURE 10.4 Container and pallet types.

Lower deck—Container

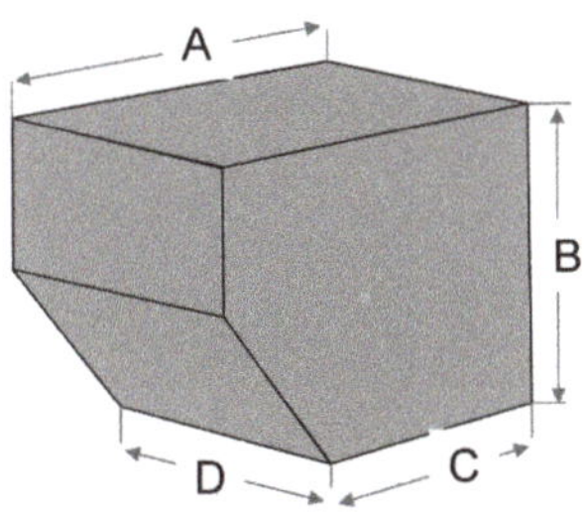

Container	A	B	C	D	Volume (ft³)
LD-1	92	64	61.5	60.4	175
LD-2	61.5	64	47	60.4	124
LD-3	79	64	61.5	60.4	159
LD-3-45	96	45	61.5	60.4	131

Lower deck pallet

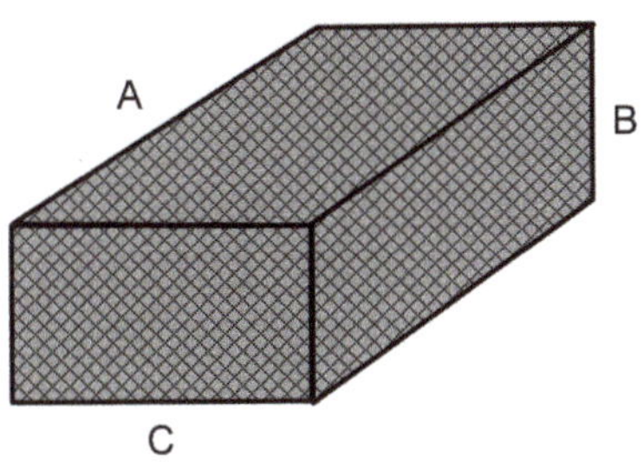

Pallet	A	B	C	Volume (ft³)
88 x 125 x 64	125	64	88	372
96 x 125 x 64	125	64	96	407

Locate the Horizontal Tail or Stabilizer in Plan View [73, 74, 78]

12. Now that we have located the wing on the fuselage, we turn our attention to the horizontal tail. In the fuselage sizing exercise, we located the aft pressure bulkhead of the cabin at a distance of 0.5 × length of the aft nonconstant section from the end of the fuselage.
13. Starting from the tail, we next define the horizontal tail pivot bulkhead. This the bulkhead that is placed at the intersection of the horizontal tail rear spar with the side of the fuselage. This bulkhead houses the horizontal tail pivot (typically located about 5 in. forward of the bulkhead) about which the horizontal tail pivots or rotates as it changes its angle of incidence.
14. The space between the horizontal tail (HT) pivot bulkhead and the fuselage tail cone accommodates what we define as the APU compartment, which houses the APU exhaust system, the APU engine, and the APU firewall.
15. The length of the APU compartment is determined by the length of the APU engine. The length of the APU engine is estimated as follows:

$$\text{APU engine length}\left(\text{in.}\right) = 0.01 \times \text{Fuselage area parameter} + 49.5 \tag{10.1}$$

$$\begin{aligned}\text{Fuselage area parameter} &= \text{Fuselage length}\left(L_f\right)\\ &\quad \times \text{Fuselage effective diameter}\left(D_f\right)\end{aligned} \tag{10.2}$$

$$\text{Fuselage effective diameter} = \text{SQRT}\left\{\text{Fuselage width}\left(W_f\right) \times \text{Fuselage height}\left(H_f\right)\right\} \tag{10.3}$$

$$\text{Length of the APU compartment} = \text{Length of the APU engine} + 75\,\text{in.} \tag{10.4}$$

Here, the 75 in. is the space required for the exhaust system and the offset distance of the APU firewall aft of the HT tail pivot bulkhead.

16. The APU firewall is offset 22 in. (±2 in.) aft of the HT pivot bulkhead. This is not shown in the figure for clarity. The primary purpose of this offset is to isolate the APU from the horizontal tail structure (and systems) should the APU experience an engine fire.
17. Once the horizontal tail is positioned, a check is made to see whether there is adequate space between the apex of the front spar and the aft pressure bulkhead.

This space is to allow for the HT jack screw, which helps change the incidence of the horizontal tail.

18. The minimum space requirement for the jack screw is needed for the mechanism itself and ease of maintenance access. For single-aisle aircraft, 24 in. works well as a minimum requirement and this value can increase up to 55 in. for a large transport such as an A380.
19. If the space available for the jack screw falls on the curve shown in the figure (±2 in.), then no additional integration work is needed. If more space is required, the first option is to seek opportunity in the APU compartment, followed by shrinking (within reason) the HT box chord at the root (by moving the front and aft spar). While not favorable, the last resort would be to push the cabin aft pressure bulkhead forward. I say last resort, because this has an impact on the cabin, which is the prime revenue generation commodity of the aircraft. Conversely, if more than adequate room is available, it should be used as an opportunity to push the aft pressure bulkhead further back to create additional cabin volume.

FIGURE 10.5 Locating the horizontal tail.

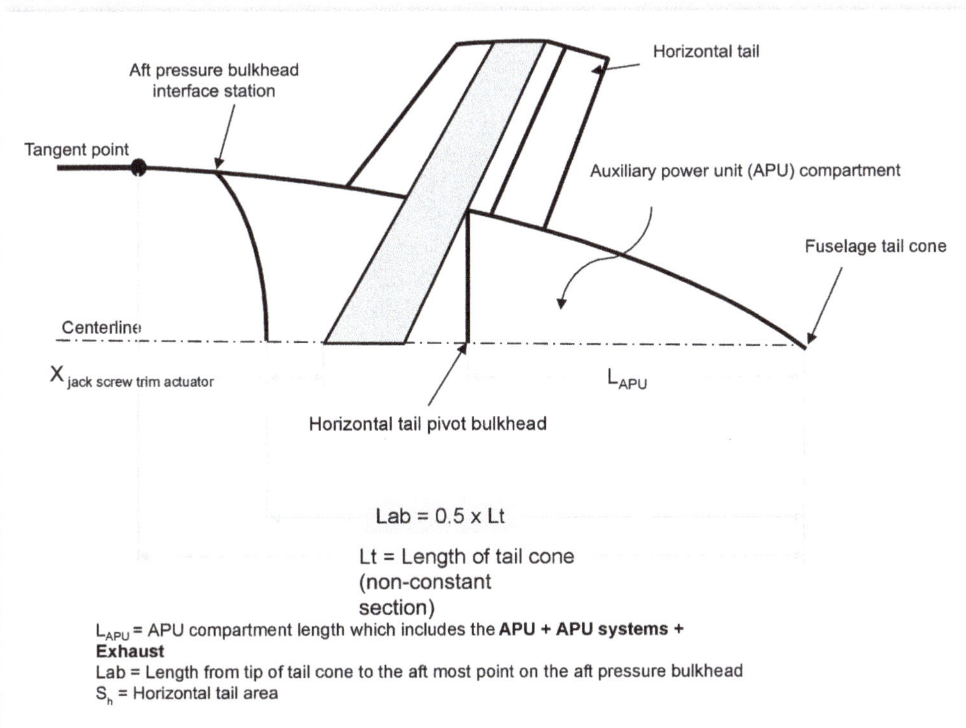

FIGURE 10.6 Sizing the jack screw and APU compartment length.

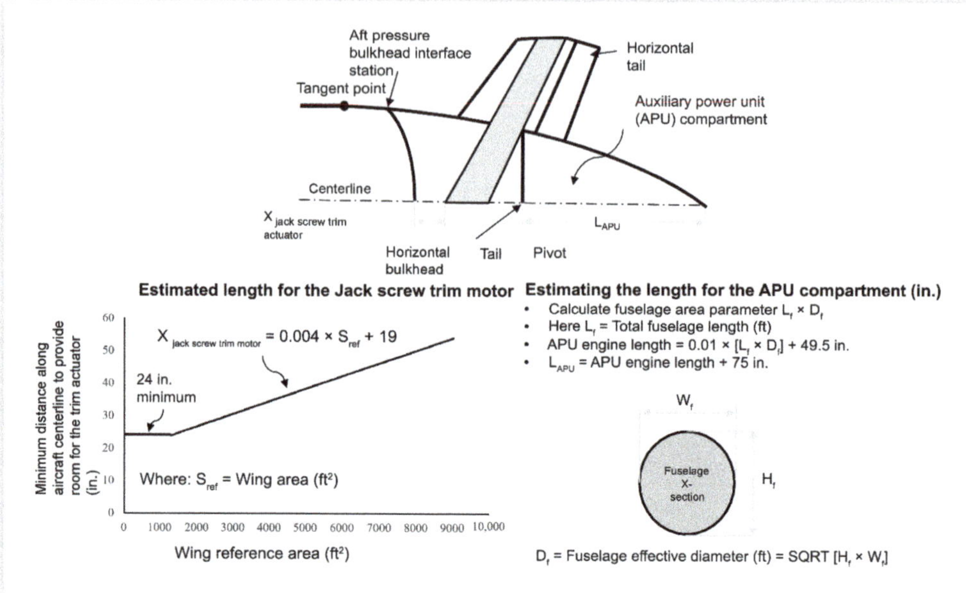

Locate the Wing in Side View [73, 74, 78]

1. Position the wing vertically on the body such that the highest point of the airfoil (typically at the front spar) is below the top of the cabin floor by approximately 6 in. (±1 in.).
2. The area above the wing typically has floor beams running longitudinally. The gap between the top of the floor and the wing allows for wiring, power feeders, and other system "transport elements" to go from the front of the aircraft to the aft end. This space or gap also serves the purpose of providing vent volume between the forward and aft fuselage during a cabin decompression event. The more volume available the better from a system integration standpoint but the lower the wing is positioned, the deeper it drives the wing-to-body fairing translating into additional weight and drag.
3. The main landing gear wheel well is positioned aft of the wing rear spar, and the upper surface of the wheel well typically forms the horizontal pressure deck. The horizontal pressure deck separates the unpressurized main landing gear wheel well from the pressurized area above it. The horizontal pressure deck is positioned such that it is 2 to 4 in. below the passenger floor.

4. The junction between the rear spar and the horizontal pressure deck is a highly loaded area, and the structure here must react to wing-bending loads, pressure loads, and deflections due to body bending. The load paths are complex, and extensive guidance is necessary from the structure's community to ensure a robust design, which is also weight-efficient.
5. The forward face of the nose wheel well shares a common bulkhead (3 to 5 in.) with the forward radome. The nose wheel well should be long enough to accommodate the length of the nose landing gear with the oleo (shock absorber) in the extended position (typical extension is 18 to 20 in.). Keep in mind the nose landing gear wheel well is an unpressurized area. The space between the nose wheel well horizontal pressure deck and the flight deck floor is determined by the room to integrate systems such as the flight deck controls and wiring that are placed below the floor.

FIGURE 10.7 Fuselage below floor inboard profile of a typical jet transport.

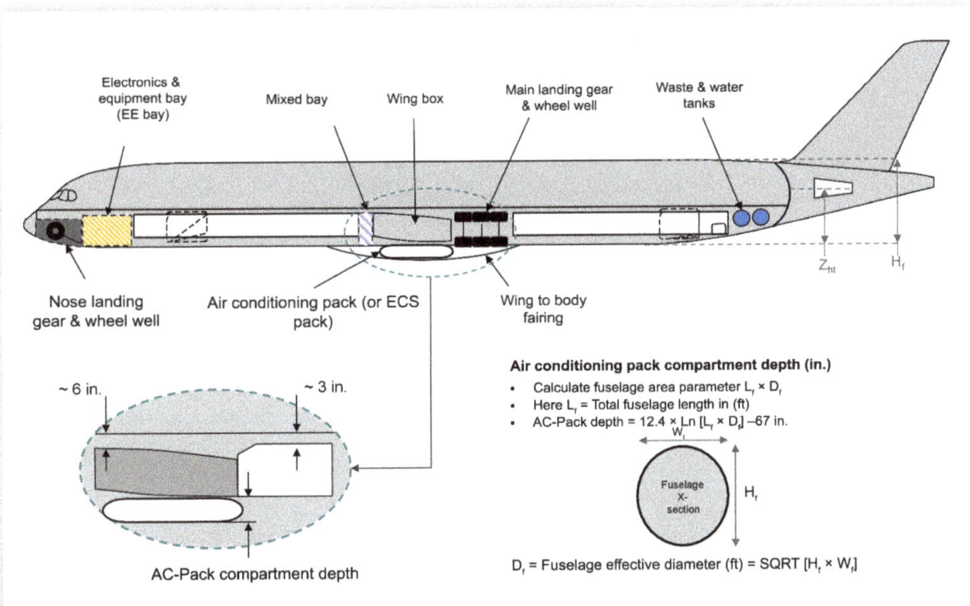

Locate the Empennage in Side View [73, 74, 78]

6. The horizontal tail is located at a distance Z_{ht} from the bottom of the fuselage (in the side view) as shown in Figure 10.8. At this early stage of design, existing airplanes are used as guidance and

$$Z_{ht} = 0.665 \times H_f \tag{10.5}$$

where
0.665 is an average value derived from existing jet transports
H_f is the height of the fuselage cross section.

7. Position the vertical tail (VT) along the body such that the front spar of the VT is approximately 22 in. ± 2 in. from the point at which the aft pressure bulkhead interfaces with a fuselage frame (as measured along the centerline of the fuselage).

Note: While not shown, the front spar of the VT typically interfaces with a frame bay.

FIGURE 10.8 Locating the vertical tail.

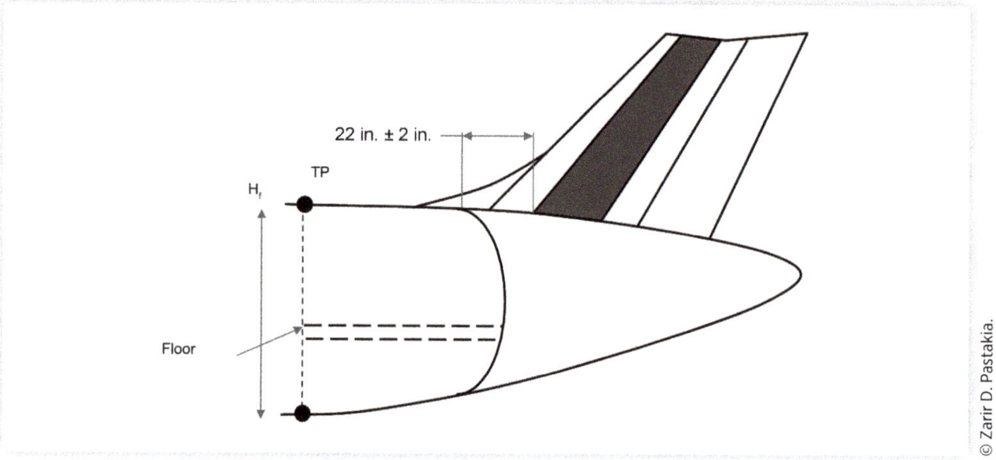

Other Considerations

- Check roll and rotation contact angles. After the final drawing is settled, a final check should be made to see whether any large deviations from the initial sizing assumptions are present. If so, some iteration will be required either in the drawing or in the sizing assumptions. The typical roll angle should be greater than 10°. This is done to prevent runway strikes during take-off and landing in windy conditions. The 747 has had its share of nacelle strikes, and its roll angle is less than 8°.
- Typical take-off and landing angles should also be verified vs. initial assumptions.

FIGURE 10.9 Roll angles and aft body contact angles.

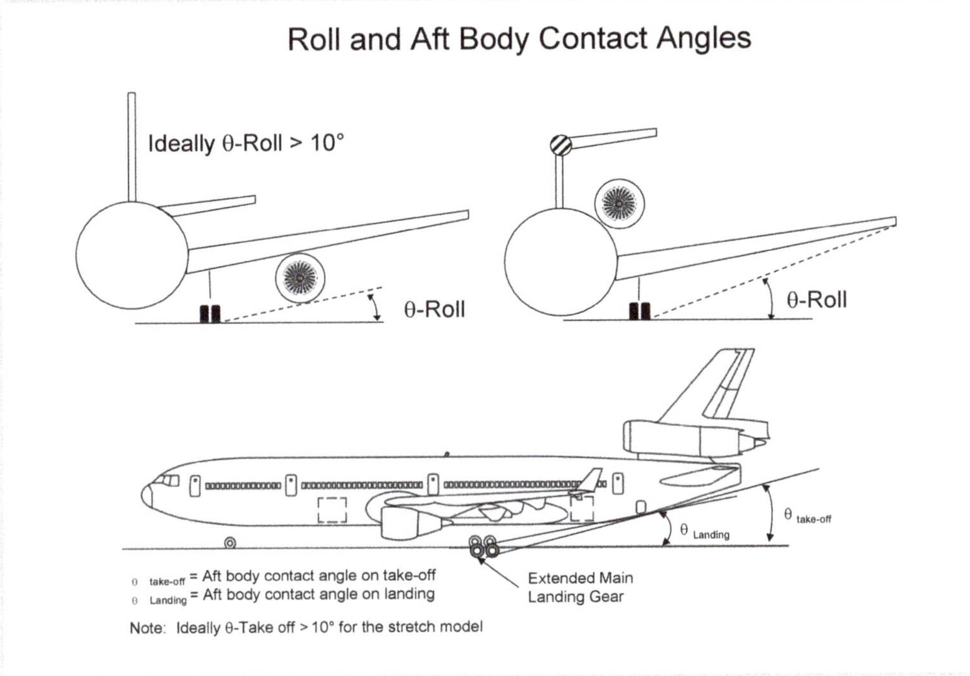

Courtesy of Boeing.

The General Arrangement or 3-View

Now that the configuration has come together, a general arrangement or three-view drawing as it is sometimes called can be created. The key elements of a general arrangement drawing are shown as an example. The drawing shown represents a future proposed aircraft drawn up by the author in the 265-seat capacity class that could replace a Boeing 767/A330 someday.

There you have it. The general arrangement can now be used to do a more detailed analysis of the configuration. The idea is to iterate on the design and adjust the drawing and the sizing until the assessed airplane meets the mission requirements.

FIGURE 10.10 General arrangement drawing (three-view).

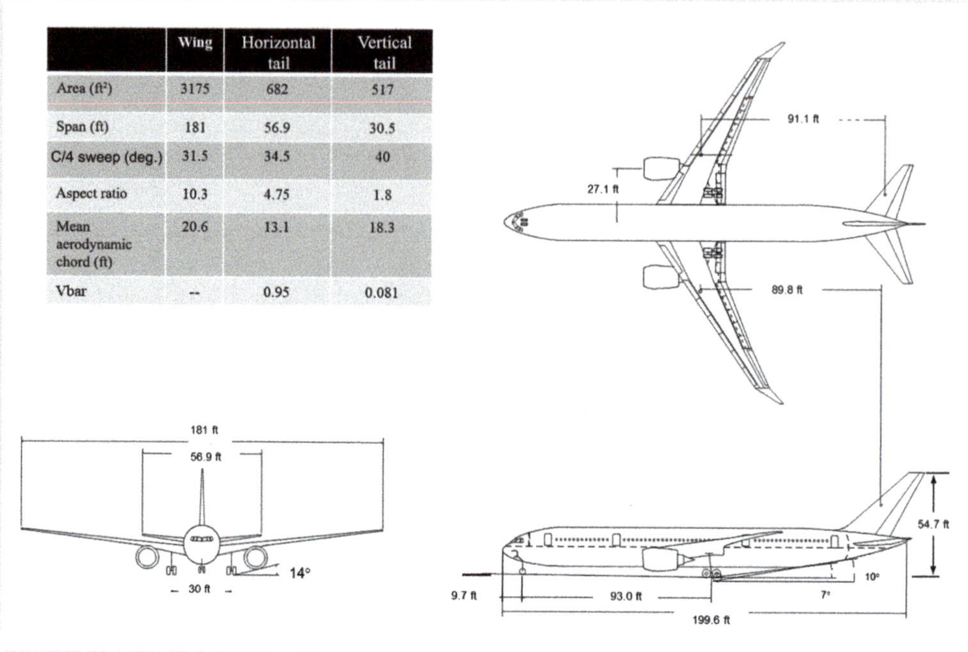

	Wing	Horizontal tail	Vertical tail
Area (ft²)	3175	682	517
Span (ft)	181	56.9	30.5
C/4 sweep (deg.)	31.5	34.5	40
Aspect ratio	10.3	4.75	1.8
Mean aerodynamic chord (ft)	20.6	13.1	18.3
Vbar	--	0.95	0.081

References

Aircraft Sizing and Design

1. Airbus, "Airport & Maintenance Planning Document," n.d.-a, www.airbus.com.
2. Boeing Company, "Airplane Characteristics for Airport Planning," n.d.-a, www.boeing.com.
3. Corning, G., *Supersonic and Subsonic Airplane Design* (Ann Arbor, MI: Edwards Brothers, 1980a).
4. Daniel, P., *Raymer, Aircraft Design: A Conceptual Approach*, 5th ed. (Washington, DC: AIAA, 2012).
5. Fielding, J., *Introduction to Aircraft Design* (New York: Cambridge University Press, 1999).
6. General Electric Corporation, n.d., www.ge.com.
7. Howe, D., *Aircraft Conceptual Design Synthesis* (London: Professional Engineering Publishing Limited, 2000).
8. Jenkinson, L., Simpkin, P., and Rhodes, D., *Civil Jet Aircraft Design* (Oxford: Butterworth-Heinemann, 1999a).
9. https://booksite.elsevier.com/9780340741528/appendices/data-a/table-1/table.htm, Website for the book, *Civil jet Aircraft Design*, by Jenkinson, Simpkin & Rhodes (Reference 8).
10. Loftin, L.K. Jr., "Subsonic Aircraft: Evolution and Matching of Size to Performance," NASA RP 1060, August 1980.
11. Nicolai, L.M., *Fundamentals of Aircraft Design* (San Jose, CA: METS, Inc., 1975).
12. Nicolai, L.M. and Carichner, G.E., *Fundamentals of Aircraft and Airship Design*, AIAA Education Series (Reston, VA: AIAA, 2010a).
13. Rolls Royce, n.d., www.rolls-royce.com.
14. Roskam, J., *Airplane Design Parts 1 through 8* (Lawrence, KS: DAR Corporation, 1989a).
15. Schaufele, R.D., *The Elements of Aircraft Preliminary Design* (Santa Ana, CA: ARIES Publications, 2000a).
16. Scholz, D., "Aircraft Design – Short Course," Hamburg University of Applied Sciences, 2007a.
17. Scott, P.W. and Nguyen, D., "The Initial Weight Estimate," SAWE Paper No. 2327, Society of Weight Engineers, June 1996.
18. Stinton, D., *The Anatomy of the Airplane* (Oxford: BSP Professional Books, 1985).
19. Stinton, D., *The Design of the Airplane* (New York: van Nostrand Reinhold, 1983).
20. Torenbeek, E., *The Synthesis of Subsonic Airplane Design* (Dordrecht, the Netherlands: Kluwer Academic Pub., 1982a).
21. Torenbeek, E., *Advanced Aircraft Design, Conceptual Design, Analysis and Optimization of Subsonic Civil Airplanes* (Chicester, UK: John Wiley & Sons, 2013).

Fuselage Design

22. Hamburg University, "'Aircraft Design—6. Fuselage Design'—An Open Educational Resources (OER) for Hamburg Open Online University (HOOU)," Hamburg University, n.d.

23. Man-Systems Integration Standards, "Anthropometry and Biomechanics," NASA-STD-3000, n.d.

24. Niu, M.C.Y., *Airframe Structural Design*, 2nd ed. (Hong Kong: Conmilit Press, Ltd., 1999a).

25. Roskam, J., *Airplane Design Parts 1 through 8* (Lawrence, KS: DAR Corporation, 1989b).

26. Schaufele, R.D., *The Elements of Aircraft Preliminary Design* (Santa Ana, CA: ARIES Publications, 2000b).

27. Schlichting, H. and Truckenbrodt, E., *Aerodynamics of the Airplane* (New York: McGraw-Hill International Book Company, 1979).

28. Stanford, "Cross-Section Design," n.d., http://aerodesign.stanford.edu/aircraftdesign/fuselayout/fusesection.html.

29. Torenbeek, E., *The Synthesis of Subsonic Airplane Design* (Dordrecht, the Netherlands: Kluwer Academic Pub., 1982b).

Wing Design

30. Corning, G., *Supersonic and Subsonic Airplane Design* (Ann Arbor, MI: Edwards Brothers, 1980b).

31. ESDU, "Geometrical Properties of Cranked and Straight Tapered Wing Planforms," ESDU 76003, n.d.

32. Kuchemann, D., *The Aerodynamic Design of Aircraft* (Oxford: Pergamon Press, 1978).

33. Mason, W.H., "Transonic Aerodynamics of Airfoils and Wings," Short Course, Virginia Tech, March 2006.

34. McCormick, B.W. Jr., *Aerodynamics, Aeronautics and Flight Mechanics*, 2nd ed. (New York: John Wiley & Sons Inc., 1995).

35. McCormick, B.W. Jr., *Aerodynamics of V/STOL Flight* (New York: Academic Press, 1967).

36. Niu, M.C.Y., *Airframe Structural Design*, 2nd ed. (Hong Kong: Conmilit Press, Ltd., 1999b).

37. Obert, E., *Aerodynamic Design of Transport Aircraft* (Amsterdam, the Netherlands: Delft University of Technology; IOS Press, 2009).

38. Phillips, W.F., Fugal, S.R., and Spall, R.E., "Minimizing Induced Drag with Geometric and Aerodynamic Twist, CFD Validation," AIAA Paper Number 2005-1034, 2005.

39. Pope, A., *Basic Wing and Airfoil Theory* (New York: McGraw-Hill Book Company Inc., 1951).

40. Rudolph, P.K.C., "High Lift Systems on Commercial Subsonic Airliners," NASA/CR 4746, n.d.

41. Scholz, D., "Aircraft Design—Short Course," Hamburg University of Applied Sciences, 2007b.

42. Torenbeek, E., *The Synthesis of Subsonic Airplane Design* (Dordrecht, the Netherlands: Kluwer Academic Pub., 1982c).

Empennage Design

43. Nicolai, L.M. and Carichner, G.E., *Fundamentals of Aircraft and Airship Design*, AIAA Education Series (Reston, VA: AIAA, 2010b).

44. Roskam, J., *Airplane Design Parts 1 through 8* (Lawrence, KS: DAR Corporation, 1989c).

45. Schaufele, R.D., *The Elements of Aircraft Preliminary Design* (Santa Ana, CA: ARIES Publications, 2000c).

46. Torenbeek, E., *The Synthesis of Subsonic Airplane Design* (Dordrecht, the Netherlands: Kluwer Academic Pub., 1982d).

Propulsion

47. Arvid, K.L. Jr., "A Brief Study of the Effects of Turbofan-Engine Bypass Ratio on Short and Long-Haul Cruise Aircraft," Langley Research Center, NASA, December 1975.

48. Daggett, D.L., Brown, S.T., and Kawai, R.T., "Ultra Efficient Engine Diameter Study," NASA/CR-2003-212309, May 2003.

49. Hunecke, K., *Jet Engines—Fundamentals of Theory, Design and Operation* (Osceola, WI: Motorbooks International, 1997).

50. Jenkinson, L.R., Simpkin, P., and Rhodes, D., *Civil Jet Aircraft Design* (Oxford: Butterworth-Heinemann, 1999c).

51. NASA, "NASA Paper on State of the Art of Subsonic Engine SFC," n.d., https://www.hq.nasa.gov/office/aero/events/encompat/soaeng.pdf.

52. Roskam, J. and Lan, C.T., *Airplane Aerodynamics and Performance* (Lawrence, KS: DAR Corporation, 2003a).

53. Svoboda, C., "Turbofan Engine Database as a Preliminary Design Tool," Department of Aerospace Engineering, The University of Kansas, March 2000.

54. Torenbeek, E., *The Synthesis of Subsonic Airplane Design* (Dordrecht, the Netherlands: Kluwer Academic Pub., 1982e).

Aircraft Performance

55. Anderson, J.D. Jr., *Aircraft Performance and Design* (Boston, MA: WCD McGraw-Hill, 1999).

56. Austyn Mair, W. and Birdsall, D.L., *Aircraft Performance* (Cambridge, UK: Cambridge University Press, 1992).

57. Filippone, A., *Advanced Aircraft Flight Performance* (Cambridge, UK: Cambridge University Press, 2012).

58. Hale, F.J., *Introduction to Aircraft Performance, Selection, and Design* (New York: John Wiley & Sons, 1984).

59. Roskam, J. and Lan, C.T., *Airplane Aerodynamics and Performance* (Lawrence, KS: DAR Corporation, 2003b).

60. Shevell, R.S., *Fundamentals of Flight*, 2nd ed. (Englewood Cliffs, NJ: Prentice Hall Inc., 1989).

61. Torenbeek, E., *Optimum Cruise Performance of Subsonic Transport Aircraft* (Delft, the Netherlands: Delft University Press, 1998).

Economics

62. Ackert, S., "Commercial Aspects of Aircraft Customization," Aircraft Monitor, 2013.

63. "Airbus New Aircraft List Prices for 2017," 2017.

64. "Boeing New Aircraft List Prices for 2017," 2017.

65. Harris, F.D., "An Economic Model of US Airline Operating Expenses," NASA Ames Research Center, December 2005.

66. Nicolai, L.M. and Carichner, G.E., *Fundamentals of Aircraft and Airship Design*, AIAA Education Series (Reston, VA: AIAA, 2010c).

67. Peters, B., "Why Boeing, Airbus Could Leave GE with LEAP Losses Near Term," *Investor's Business Daily*, June 7, 2016.

68. Roskam, J., *Airplane Design Parts 1 through 8* (Lawrence, KS: DAR Corporation, 1989d).

69. Simple Flying, "'Simple Flying' Article on Airbus Pricing," February 18, 2019, https://simpleflying.com/airbus-reveals-on-average-airlines-get-50-off-airline-list-prices/.

70. Torenbeek, E., *The Synthesis of Subsonic Airplane Design* (Dordrecht, the Netherlands: Kluwer Academic Pub., 1982f).

71. Wilcox, K., "16.885J Aircraft Systems Engineering Cost Analysis," Lecture Notes, Aerospace Computational Design Laboratory, MIT, September 19, 2004.

Other

72. Airbus, "Airport & Maintenance Planning Document," n.d., https://www.airbus.com.

73. Boeing Company, "Airplane Characteristics for Airport Planning," n.d., https://www.Boeing.com.

74. Brooks, R.R., *The Handley Page Victor—The History and Development of a Classic Jet*. Vol. 1 (Barnsley, UK: Pen & Sword Aviation, 2007).

75. Civil Aviation Authority, "Quota Count Validation Study at Heathrow Airport," Civil Aviation Authority, UK, 2020.

76. Dickson, N., "Progress on the Development of the CO_2 Standard," ICAO Environment Branch, May 2013.

77. Embraer, "Airport Planning Manual," E190, n.d., https://www.Embraer.com.

78. Feagin, R.C. and Morrison, W.D. Jr., "Delta Method, an Empirical Drag Buildup Technique," Contract NAS 2-8612, Lockheed California Company, December 1978.

79. Fields, D.W., *The Avro Type 698 Vulcan —The Secrets Behind Its Design and Development* (Barnsley, UK: Pen & Sword Aviation, 2012).

80. International Industry Working Group, "Commercial Aircraft Design Characteristics—Trends and Growth Projections," Fifth Edition R1, International Industry Working Group, 2007.

81. Michelin Tire Company, "Michelin Commercial—Regional Application Chart for Aviation Tires," n.d.

82. MRAeS, P.J., *Janes All the Worlds Aircraft* (Alexandria, VA: Janes Information Group Inc., 2004-2005).

83. Stephen Arrow Smith (European Aviation Safety Agency) & Laszlo Wind Hoeffer (US Federal Aviation Administration), "The CAEP/10 Recommendation on a New ICAO Aeroplane CO_2 Emissions Standard," ICAO Environmental Report, 2016, 112-114.

84. Wake Vortices, "Safety First," *Airbus Magazine*, January 2016.

85. Wikipedia, January 21, 2021, https://www.Wikipedia.com.

Appendix A: The Standard Atmosphere

Altitude (ft)	Temperature, R (Rankine)	Pressure, p (psi)	Density, ρ (slugs/ft^3)	Temperature ratio, θ	Pressure ratio, p/p_0	Density ratio, ρ/ρ_0	Coefficient of dynamic viscosity, μ (slugs/ft-sec)	Speed of sound (fps)	Speed of sound (knots)
0	518.7	14.7	0.002377	1	1	1	3.745E−07	1116.4	661.4
1000	515.1	14.1765	0.002308	0.9931	0.9644	0.9711	3.724E−07	1112.5	659.2
2000	511.6	13.6682	0.002241	0.9863	0.9298	0.9428	3.704E−07	1108.7	656.9
3000	508.0	13.1748	0.002175	0.9794	0.8962	0.9151	3.683E−07	1104.8	654.6
4000	504.4	12.6959	0.002111	0.9725	0.8637	0.8881	3.663E−07	1100.9	652.3
5000	500.9	12.2312	0.002048	0.9656	0.8321	0.8617	3.642E−07	1097.0	650.0
6000	497.3	11.7804	0.001987	0.9588	0.8014	0.8359	3.621E−07	1093.1	647.7
7000	493.7	11.3431	0.001927	0.9519	0.7716	0.8107	3.600E−07	1089.2	645.3
8000	490.2	10.9190	0.001868	0.9450	0.7428	0.7860	3.579E−07	1085.3	643.0
9000	486.6	10.5079	0.001811	0.9381	0.7148	0.7620	3.558E−07	1081.3	640.6
10,000	483.0	10.1094	0.001755	0.9313	0.6877	0.7385	3.537E−07	1077.3	638.3
11,000	479.5	9.7233	0.001701	0.9244	0.6614	0.7156	3.516E−07	1073.3	635.9
12,000	475.9	9.3491	0.001648	0.9175	0.6360	0.6932	3.495E−07	1069.3	633.6
13,000	472.3	8.9867	0.001596	0.9106	0.6113	0.6713	3.473E−07	1065.3	631.2
14,000	468.8	8.6358	0.001545	0.9038	0.5875	0.6500	3.452E−07	1061.3	628.8
15,000	465.2	8.2961	0.001496	0.8969	0.5644	0.6292	3.431E−07	1057.3	626.4
16,000	461.6	7.9672	0.001447	0.8900	0.5420	0.6090	3.409E−07	1053.2	624.0
17,000	458.1	7.6490	0.001400	0.8831	0.5203	0.5892	3.387E−07	1049.1	621.6
18,000	454.5	7.3412	0.001355	0.8763	0.4994	0.5699	3.366E−07	1045.0	619.2
19,000	450.9	7.0434	0.001310	0.8694	0.4791	0.5511	3.344E−07	1040.9	616.7
20,000	447.4	6.7556	0.001266	0.8625	0.4596	0.5328	3.322E−07	1036.8	614.3
21,000	443.8	6.4773	0.001224	0.8556	0.4406	0.5150	3.300E−07	1032.7	611.8
22,000	440.2	6.2084	0.001183	0.8488	0.4223	0.4976	3.278E−07	1028.5	609.4
23,000	436.7	5.9486	0.001143	0.8419	0.4047	0.4807	3.256E−07	1024.3	606.9
24,000	433.1	5.6976	0.001103	0.8350	0.3876	0.4642	3.234E−07	1020.1	604.4
25,000	429.5	5.4553	0.001065	0.8281	0.3711	0.4481	3.212E−07	1015.9	601.9
26,000	426.0	5.2215	0.001028	0.8213	0.3552	0.4325	3.190E−07	1011.7	599.4
27,000	422.4	4.9958	0.000992	0.8144	0.3398	0.4173	3.167E−07	1007.5	596.9
28,000	418.9	4.7780	0.000957	0.8075	0.3250	0.4025	3.145E−07	1003.2	594.4
29,000	415.3	4.5681	0.000923	0.8006	0.3108	0.3881	3.122E−07	998.9	591.8
30,000	411.7	4.3656	0.000889	0.7938	0.2970	0.3741	3.100E−07	994.6	589.3
31,000	408.2	4.1705	0.000857	0.7869	0.2837	0.3605	3.077E−07	990.3	586.7
32,000	404.6	3.9825	0.000826	0.7800	0.2709	0.3473	3.054E−07	986.0	584.2
33,000	401.0	3.8014	0.000795	0.7731	0.2586	0.3345	3.031E−07	981.6	581.6

Altitude (ft)	Temperature, R (Rankine)	Pressure, p (psi)	Density, ρ (slugs/ft³)	Temperature ratio, θ	Pressure ratio, p/p_0	Density ratio, ρ/ρ_0	Coefficient of dynamic viscosity, μ (slugs/ft-sec)	Speed of sound (fps)	Speed of sound (knots)
34,000	397.5	3.6271	0.000765	0.7663	0.2467	0.3220	3.008E-07	977.2	579.0
35,000	393.9	3.4593	0.000737	0.7594	0.2353	0.3099	2.985E-07	972.8	576.4
36,000	390.3	3.2978	0.000709	0.7525	0.2243	0.2981	2.962E-07	968.4	573.8
36,089	390.0	3.2837	0.000706	0.7519	0.2234	0.2971	2.960E-07	968.0	573.5
37,000	390.0	3.1433	0.000676	0.7519	0.2138	0.2844	2.960E-07	968.0	573.5
38,000	390.0	2.9958	0.000644	0.7519	0.2038	0.2710	2.960E-07	968.0	573.5
39,000	390.0	2.8553	0.000614	0.7519	0.1942	0.2583	2.960E-07	968.0	573.5
40,000	390.0	2.7213	0.000585	0.7519	0.1851	0.2462	2.960E-07	968.0	573.5
41,000	390.0	2.5936	0.000558	0.7519	0.1764	0.2346	2.960E-07	968.0	573.5
42,000	390.0	2.4719	0.000532	0.7519	0.1682	0.2236	2.960E-07	968.0	573.5
43,000	390.0	2.3559	0.000507	0.7519	0.1603	0.2131	2.960E-07	968.0	573.5
44,000	390.0	2.2454	0.000483	0.7519	0.1527	0.2031	2.960E-07	968.0	573.5
45,000	390.0	2.1400	0.000460	0.7519	0.1456	0.1936	2.960E-07	968.0	573.5
46,000	390.0	2.0396	0.000439	0.7519	0.1388	0.1845	2.960E-07	968.0	573.5
47,000	390.0	1.9439	0.000418	0.7519	0.1322	0.1759	2.960E-07	968.0	573.5
48,000	390.0	1.8527	0.000398	0.7519	0.1260	0.1676	2.960E-07	968.0	573.5
49,000	390.0	1.7658	0.000380	0.7519	0.1201	0.1598	2.960E-07	968.0	573.5
50,000	390.0	1.6829	0.000362	0.7519	0.1145	0.1523	2.960E-07	968.0	573.5
55,000	390.0	1.3235	0.000285	0.7519	0.0900	0.1197	2.960E-07	968.0	573.5
60,000	390.0	1.0408	0.000224	0.7519	0.0708	0.0942	2.960E-07	968.0	573.5
65,000	390.0	0.8185	0.000176	0.7519	0.0557	0.0740	2.960E-07	968.0	573.5
70,000	390.0	0.6437	0.000138	0.7519	0.0438	0.0582	2.960E-07	968.0	573.5

Note: The shaded row is the tropopause, above which the temperature remains constant.

Appendix B: Different Approaches to the Same Requirement. There Is No Substitute for Creativity and Imagination

In 1947, the British Air Ministry put out specification B35/46 to support operational requirement OR229. The requirement was for a "four-engine swept-wing jet medium bomber with a cruising speed of 500 kt and a ceiling of at least 55,000 ft" that could deliver a nuclear bomb. The requirement was in response to the growing threat from the Soviet Union. Jet bombers are the closest military cousin to transonic transport jets from a design standpoint, given the need to cruise long distances at high transonic speed. Hence, the usual aerodynamic metrics like good lift-to-drag ratio are important, but so are safe high-speed maneuvering capability which requires an adequate usable lift coefficient to avoid buffet.

Specification B35/46—Summary of Aircraft Requirements [65, 80]

- Key armament: 10,000-lb nuclear bomb, 60-in. diameter and 290-in. length.
- Radius of action: 1500 NMI, or equivalent still air range of 3350 NMI carrying payload of 10,000 lb to and from target (assuming mission is aborted).
- Take-off field length: Take-off in tropical conditions, 4500 to 50 ft screen height (rocket-assisted take-off acceptable). Later this requirement was relaxed to 7500 ft under normal take-off power (w/o rocket assistance).
- Approach speed: 120 knots.
- Maneuvering at high speed: Maximum Mach 0.90 maneuvering capability during combat with safe flying characteristics.
- Engine failure in cruise: Minimum cruising height of 45,000 ft with one engine inoperative. Minimum cruising height of 30,000 ft with two engines inoperative.

These were extremely stringent requirements for the time and demanded significant leaps in aerodynamic, systems, airframe, and propulsion technology. The Avro Vulcan and Handley Page Victor are great examples of how two different design teams approached and satisfied the requirements, with very different designs. The Victor clearly offered a better lift-to-drag ratio over the Vulcan, but both aircraft ended up succeeding in their intended roles.

With the advent of the surface to air missile, it became clear that the 50,000-ft cruise altitude requirement was not adequate to keep the bomber out of harm's way, and the mission was modified for an initial cruise to target at high altitude, followed by a final run to the target at low altitude. It was soon discovered that low-altitude flying (in practice) led to cracks in the Victor's wing due to fatigue. Fatigue was likely caused by the susceptibility of the Victor's higher aspect ratio (and associated higher lift curve slope) wing to increased maneuvering loads in the turbulent air at lower altitudes. In the end, the Vulcan proved to be a more versatile design for the modified mission; however, the Victor found renewed life as a high speed, transonic tanker.

Interestingly, both were used successfully in the Falkland Islands War, where 11 Victor tankers were utilized to refuel two Vulcan bombers across the South Atlantic ocean, to execute a strike against the British islands, which were being occupied by Argentinian forces. This is a great example of two very different aircraft configuration designs in response to the same requirement.

FIGURE B.1 The Handley page victor and the Avro Vulcan.

The Handley Page Victor & Avro Vulcan

Different designs in response to Spec. B35/46, operating requirement OR 229, in 1947

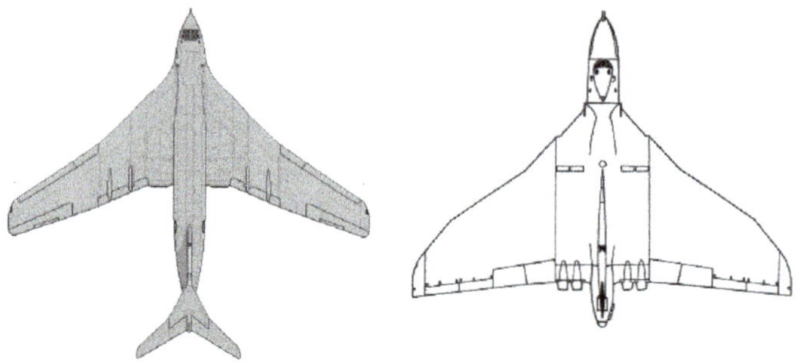

Handley Page Victor Mk 1		Vulcan B. Mk 2
2406	Wing Area S (ft^2)	3964
110	Wingspan b (ft)	111
5.0	Aspect Ratio AR	3.1
77	Max Wing Loading (W/S) psf	53.0
102.4	Fuselage length (ft)	97.1
8400	Awet (ft^2) est.	10550
0.02070	Total Drag Coeff. CD est.	0.02060
50	CD. S	82
0.385	CL cruise	0.355
18.6	L/D max @ M. L/Dmax	17.2

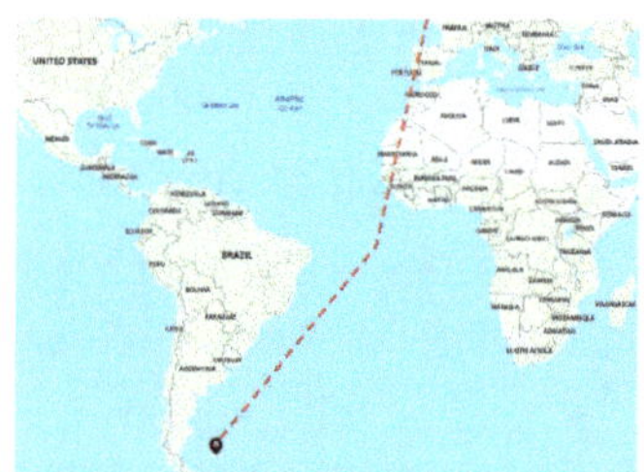

Appendix C: Rapid Estimating Methods

Effect of Aspect Ratio on the Drag Polar

The drag polar of an aircraft can be scaled based on changes in the wing aspect ratio as follows.

$$CD_2 = CD_1 + CL^2 \times \{1/(\pi \times AR_2) - 1/(\pi \times AR_1)\} \quad \text{(C.1)}$$

where
1 is the known value from existing aircraft
2 is the new aircraft value
CD is the total drag coefficient
AR is the aspect ratio

This sort of rapid estimation is useful for "what-if" studies, especially when considering a derivative aircraft, that could benefit from a small span extension, such as when going from the A320ceo to an A320ceo with a sharklet, or a B777-300 to a B777-300ER.

Propulsion Scaling [60]

For a given bypass ratio, engine dimensions can be scaled using a scaling factor (SF):

$$SF = Thrust_{required} / Thrust_{data} \quad \text{(C.2)}$$

$$L_{req} = L_{data} \times (SF)^{0.4} \quad \text{(C.3)}$$

$$D_{req} = D_{data} \times (SF)^{0.5} \quad \text{(C.4)}$$

$$W_{req} = W_{data} \times (SF)^{1.1} \quad \text{(C.5)}$$

where
L is the length of the engine
D is the fan diameter of the engine
W is the weight of the engine

These scaling methods are useful when using sizing tools to "rubberize" an engine based on thrust. These methods work well given their simplicity, and they can sometimes be useful to gain insights into the capability of a product. A case in point is the series of geared turbofan engines from Pratt & Whitney. If you start with the PW1100G, it has a thrust of 33,100 lbf and a fan diameter of 81 in. You can use the scalars to solve for thrust or for geometry for engines with the same bypass ratio. The PW1500G has an advertised thrust of 24,400 lbf and a fan diameter of 73 in. It has the same bypass ratio as the PW1100. Scaling the engine at the given fan diameter would, however, indicate that all else being equal, the engine might be capable of over 26,800 lbf of thrust given its 73-in. fan diameter. This insight being that there may be thrust growth potential in this engine and, consequently, the airframe. Interestingly, there have been several news reports of a stretched A220-500, which could use just that thrust growth capability.

This is one example of how such simple methods could be used to be better informed about the capability a product line could offer.

Appendix D: Data on Existing Jet Transports

- E190-E2
- A321neo
- A330-300
- B787-9
- A350-900
- B777-300ER
- MD-11
- B747-8
- A380-800

FIGURE D.1 Embraer 190-E2.

Fasttailwind/Shutterstock.com.

Embraer 190-E2 Jet

Manufacturer:	Embraer, Brazil
Entry Into Service:	2018
Engine:	PW 1900G Family
Bypass Ratio:	12
Fan Diameter:	73 in.
Thrust (0.977 × Sea Level Static Thrust):	22,000 lbf (per engine)
Likely Growth:	26,000 lbf (per engine)
Wing Area:	1109 ft^2
Wingspan	
(Geometric):	110 ft 8 in.
(Effective Aerodynamic):	Same
Wing Sweep at quarter chord:	24° est.
Total Fuel Capacity at 6.7 lb/gal:	4505 gal
High Lift System:	
Trailing Edge:	SS inbd. + SS outbd.
Leading Edge:	Slat inbd. + Slat outbd.
Fuselage Length:	119 ft 2 in.
X-Section Height:	132 in.
X-Section Width:	118.5 in.
Nose Landing Gear Tires:	Two 24 × 7.7 16PR
Main Landing Gear Tires:	Four H41 × 16-20 22PR

Design Weights

Max. Ramp Weight:	124,781 lb
Max Take-Off Weight:	124,340 lb
Max. Landing Weight:	108,136 lb
Max Zero Fuel Weight:	102,955 lb
Estimated Book Method (using above weights)	
Passenger 2 Class:	88 at 225 lb/pax
Range (200 NMI Alternate):	3235 NMI

The aircraft is fuel volume limited at a take-off weight of 123,300 lb

TOFL ISA+15°C at MTOW:	6000 ft

FIGURE D.2 A321-XLR.

Wirestock Creators/Shutterstock.com.

Airbus A321neo—XLR

Manufacturer:	Airbus Industries, Europe
Entry Into Service:	2023 est.
Engine/BPR/Fan Dia.:	PW 1100G/12/81 in.
	CFMLEAP 1A/11/78 in.
Thrust: (0.977 × Sea Level Static Thrust):	32,200 lbf (per engine)
Wing Area:	1350 ft^2 est.
Wingspan	
(Geometric):	117 ft 5 in.
(Effective Aerodynamic):	126 ft 4 in. est.
Wing Sweep at quarter chord:	25°
Total Fuel Capacity at 6.7 lb/gal:	
(With body tank):	(6260 + 3475) gal estimated.
High Lift System:	
Trailing Edge:	SS inbd. + DS outbd.
Leading Edge:	Slat inbd. + Slat outbd.
Fuselage Length:	142 ft 0 in.
X-Section Height:	163 in.
X-Section Width:	155 in.
Nose Landing Gear Tires:	Two 30 × 8.8 15PR
Main Landing Gear Tires:	Four 49 × 18-22PR

Design Weights

Max. Ramp Weight:	223,500 lb estimated.
Max Take-Off Weight:	223,000 lb estimated.
Max. Landing Weight:	179,000 lb estimated.
Max Zero Fuel Weight:	169,000 lb estimated.
Estimated Book Method (using above weights)	
Passenger 2 Class:	193 at 225 lb/pax
Range (200 NMI Alternate):	4335 NMI
TOFL ISA+15°C at MTOW:	9550 ft

Note: At the time of the writing of this book, the A321neo-XLR is in the process of revising its integral body tank, which will lead to a decrease in fuel volume and range as a result of a small increase in empty weight. The purpose of these changes was to meet crashworthiness requirements as interpreted by the regulators.

FIGURE D.3 A330-300.

JEERAPAN JANKAEW/Shutterstock.com.

Airbus A330-300

Manufacturer:	Airbus Industries, Europe
Entry Into Service:	1994
Engine/BPR/Fan Dia.:	RR T700/~5.5/97.4 in.
Thrust: (0.977 × Sea Level Static Thrust):	71,100 lbf (per engine)
Wing Area:	3850 ft^2
Wingspan	
(Geometric):	197 ft 10 in.
(Effective Aerodynamic):	198 ft 3 in. est.
Wing Sweep at quarter chord:	29.7°
Total Fuel Capacity at 6.7 lb/gal:	(36,744 gal)
High Lift System:	
Trailing Edge:	SS inbd. + SS outbd.
Leading Edge:	Slat inbd. + Slat outbd.
Fuselage Length:	206 ft 4 in.
X-Section Height:	222 in.
X-Section Width:	222 in.
Nose Landing Gear Tires:	Two 1050 × 395 R16 mm
Main Landing Gear Tires:	Eight 54 × 21 23PR in.

Design Weights

Max. Ramp Weight:	535,100 lb
Max Take-Off Weight:	533,600 lb
Max. Landing Weight:	412,335 lb
Max Zero Fuel Weight:	377,055 lb
Estimated Book Method (using above weights)	
Passenger 2 Class:	287 at 225 lb/pax
Range (200 NMI Alternate):	5815 NMI
TOFL ISA+15°C at MTOW:	10,300 ft

FIGURE D.4 B787-9.

Boeing 787-9 Dreamliner

Manufacturer:	The Boeing Company, US
Entry Into Service:	2014
Engine/BPR/Fan Dia.:	RR T1000/~9/111 in.
	GEnX/~9/111 in.
Thrust: (0.977 × Sea Level Static Thrust):	76,100 lbf (per engine)
Wing Area:	3880 ft^2
Wingspan	
(Geometric):	197 ft 5 in.
(Effective Aerodynamic):	Same
Wing Sweep at quarter chord:	32.5° est.
Total Fuel Capacity at 6.7 lb/gal:	33,380 gal
High Lift System:	
Trailing Edge:	SS inbd. + Flaperon + SS outbd. (With spoiler droop)
Leading Edge:	Slat inbd. + Sealed Krueger. + Slat outbd.
Fuselage Length:	203 ft 5 in.
X-Section Height:	226.5 in.
X-Section Width:	234.5 in.
Nose Landing Gear Tires:	Two 40 × 16 R16 26PR
Main Landing Gear Tires:	Eight 54 × 21 R23 38PR in.

Design Weights

Max. Ramp Weight:	561,500 lb
Max Take-Off Weight:	560,000 lb
Max. Landing Weight:	425,000 lb
Max Zero Fuel Weight:	400,000 lb
Estimated Book Method (using above weights)	
Passenger 2 Class:	293 at 225 lb/pax
Range (200 NMI Alternate):	7550 NMI
TOFL ISA+15°C at MTOW:	9825 ft

FIGURE D.5 Airbus A350-900 XWB.

Benson Truong/Shutterstock.com.

A350-900 XWB

Manufacturer:	Airbus Industries, Europe
Entry Into Service:	2014
Engine/BPR/Fan Dia.:	RR Trent-XWB/~9/111 in.
Thrust: (0.977 × Sea Level Static Thrust):	84,000 lbf (per engine)
Wing Area:	4640 ft^2 est.
Wingspan	
(Geometric):	212 ft 5 in.
(Effective Aerodynamic):	217 ft 11 in. est.
Wing Sweep at quarter chord:	33° est.
Total Fuel Capacity at 6.7 lb/gal:	43,589 gal
High Lift System:	
Trailing Edge:	SS inbd. + SS outbd. (with spoiler droop)
Leading Edge:	Droop Nose inbd. + Slat outbd.
Fuselage Length:	214 ft 1 in.
X-Section Height:	234.6 in.
X-Section Width:	239.8 in.
Nose Landing Gear Tires:	Two 1050 × 395R16 28PR
Main Landing Gear Tires:	Eight 1400 × 530R23 42PR

Design Weights

Max. Ramp Weight:	619,279 lb
Max Take-Off Weight:	617,400 lb
Max. Landing Weight:	452,000 lb
Max Zero Fuel Weight:	423,360 lb
Estimated Book Method (using above weights)	
Passenger 2 Class:	319 at 225 lb/pax
Range (200 NMI Alternate):	7860 NMI
TOFL ISA+15°C at MTOW:	9675 ft

FIGURE D.6 Boeing 777-300ER.

Bradley Caslin/Shutterstock.com.

Boeing 777-300ER

Manufacturer:	The Boeing Company, US
Entry Into Service:	2004
Engine/BPR/Fan Dia.:	GE90-115B/~7/128 in.
Thrust: (0.977 × Sea Level Static Thrust):	115,300 lbf (per engine)
Wing Area:	4700 ft^2 est.
Wingspan	
(Geometric):	212 ft 7 in.
(Effective Aerodynamic):	Same
Wing Sweep at quarter chord:	31.6°
Total Fuel Capacity at 6.7 lb/gal:	47,890 gal
High Lift System:	
Trailing Edge:	DS inbd.+ Flaperon + SS outbd.
Leading Edge:	Slat inbd. + Krueger + Slat outbd.
Fuselage Length:	239 ft 9 in.
X-Section Height:	244 in.
X-Section Width:	244 in.
Nose Landing Gear Tires:	Two 43 × 17.5 R17
Main Landing Gear Tires:	Eight 52 × 21.0R22

Design Weights

Max. Ramp Weight:	777,000 lb
Max Take-Off Weight:	775,000 lb
Max. Landing Weight:	554,000 lb
Max Zero Fuel Weight:	524,000 lb (opt. 529,000 lb)
Estimated Book Method (using above weights)	
Passenger 2 Class:	384 at 225 lb/pax
Range (200 NMI Alternate):	7437 NMI
TOFL ISA+15°C at MTOW:	10,550 ft

FIGURE D.7 A FedEx cargo version of the MD-11, displaying the iconic lines of this unique tri-jet airliner.

Trevisan Aviation Images/Alamy Stock Photo.

Boeing (McDonnell Douglas) MD-11

Manufacturer:	McDonnell Douglas (Now the Boeing Company, US)
Entry Into Service:	1990
Engine/BPR/Fan Dia.:	CF6-80C2/~5.5/93 in.
	PW4460-62/~5.0/94 in.
Thrust: (0.977 × Sea Level Static Thrust):	62,000 lbf (per engine)
Wing Area:	3648 ft^2 est.
Wingspan	
(Geometric):	170 ft 6 in.
(Effective Aerodynamic):	176 ft est.
Wing Sweep at quarter chord:	35°
Total Fuel Capacity at 6.7 lb/gal:	41,615 gal
High Lift System:	
Trailing Edge:	DS inbd. + DS outbd.
Leading Edge:	Slat inbd. + Slat outbd.
Fuselage Length:	192 ft 5 in.
X-Section Height:	237 in.
X-Section Width:	237 in.
Nose Landing Gear Tires:	Two 40 × 15.5 – 16
Main Landing Gear Tires:	Ten H54 × 21.0 – 24

Design Weights

Max. Ramp Weight:	633,000 lb
Max Take-Off Weight:	630,500 lb
Max. Landing Weight:	430,000 lb
Max Zero Fuel Weight:	400,000 lb

Estimated Book Method (using above weights)

Passenger 2 Class:	272 at 225 lb/pax
Range (200 NMI Alternate):	7105 NMI
TOFL ISA+15°C at MTOW:	10,800 ft

FIGURE D.8 Boeing 747-8 Intercontinental.

EQRoy/Shutterstock.com.

Boeing 747-8 Intercontinental

Manufacturer:	The Boeing Company, US
Entry Into Service:	2011 (747-8F) -2012 (747-8i)
Engine/BPR/Fan Dia.:	GEnX-B2/~8.6/104.5 in.
Thrust: (0.977 × Sea Level Static Thrust):	66,500 lbf (per engine)
Wing Area:	5960 ft^2 est.
Wingspan	
(Geometric):	224 ft 5 in.
(Effective Aerodynamic):	Same
Wing Sweep at quarter chord:	37.5°
Total Fuel Capacity at 6.7 lb/gal:	63,034 gal
High Lift System:	
Trailing Edge:	DS inbd. + Flaperon + SS outbd.
Leading Edge:	Rigid Krueger Flap inboard + Variable Camber Krueger outbd.
Fuselage Length:	243 ft 5 in.
X-Section Height:	256 in.
X-Section Width:	309 in.
Nose Landing Gear Tires:	Two 50 × 20 R22 26PR
Main Landing Gear Tires:	Sixteen 52 × 21 R22 36PR

Design Weights

Max. Ramp Weight:	990,000 lb
Max Take-Off Weight:	987,000 lb
Max. Landing Weight:	688,000 lb
Max Zero Fuel Weight:	651,000 lb

Estimated Book Method (using above weights)

Passenger 3 Class, Main + Upper Deck:	421 = 399 + 22 at 225 lb/pax
Range (200 NMI Alternate):	7735 NMI
TOFL ISA+15°C MTOW:	10,600 ft

FIGURE D.9 Airbus A380-800.

Philip Pilosian/Shutterstock.com.

Airbus A380

Manufacturer:	Airbus, Europe
Entry Into Service:	2007
Engine/BPR/Fan Dia.:	RR T900/~8.6/104.5 in.
Thrust: (0.977 × Sea Level Static Thrust):	81,000 lbf (per engine)
Wing Area:	9010 ft^2 est.
Wingspan	
(Geometric):	261 ft 8 in.
(Effective Aerodynamic):	271 ft 4 in. est.
Wing Sweep at quarter chord:	33.5°
Fuel Capacity at 6.7 lb/gal:	63,034 gal
High Lift System:	
Trailing Edge:	SS inbd. + SS outbd. + aileron droop
Leading Edge:	Droop Flap inboard + Slats outbd.
Fuselage Length:	230 ft 11.6 in.
X-Section Height:	281 in.
X-Section Width:	331 in.
Nose Landing Gear Tires:	Two 50 × 20 R22 34PR
Main Landing Gear Tires:	Twenty 1400 × 530 R23 40PR

Design Weights

Max. Ramp Weight:	1,272,100 lb
Max Take-Off Weight:	1,267,700 lb
Max. Landing Weight:	870,800 lb
Max Zero Fuel Weight:	813,500 lb

Estimated Book Method (using above weights)

Passenger 3 Class, Main + Upper Deck:	530 = 408 + 122 at 225 lb/pax
Range (200 NMI Alternate):	7760 NMI
TOFL ISA+15°C at MTOW:	11,000 ft

Appendix E: Environmental Factors

Noise [76, 77]

Definitions

The International Civil Aviation Organization (ICAO) is a specialized agency of the United Nations. It changes the principles and techniques of international air navigation and fosters the planning and development of international air transport to ensure safe and orderly growth. Its headquarters is in the Quartier International of Montreal, Quebec, Canada.

The Committee on Aviation Environment Protection (CAEP) is a technical committee of the ICAO Council established in 1983. CAEP assists the Council in formulating new policies and adopting new Standards and Recommended Practices (SARPs) related to aircraft noise and emissions, and more generally to aviation environmental impact.

Cumulative levels are defined as the arithmetic sum of the certification levels at each of the three points.

The certification procedure, specified in Chapter 3 of ICAO Annex 169, requires the determination of arrival and departure EPNLs. Three reference measurement points are specified: approach, under a 3°-descent path 2 km from the runway threshold; lateral, 450 m to the side of the initial climb after take-off, at the longitudinal position where noise is greatest; and flyover, under the departure climb path, 6.5 km from start-of-roll (SOR).

FIGURE E.1 Aircraft noise reference points.[1]

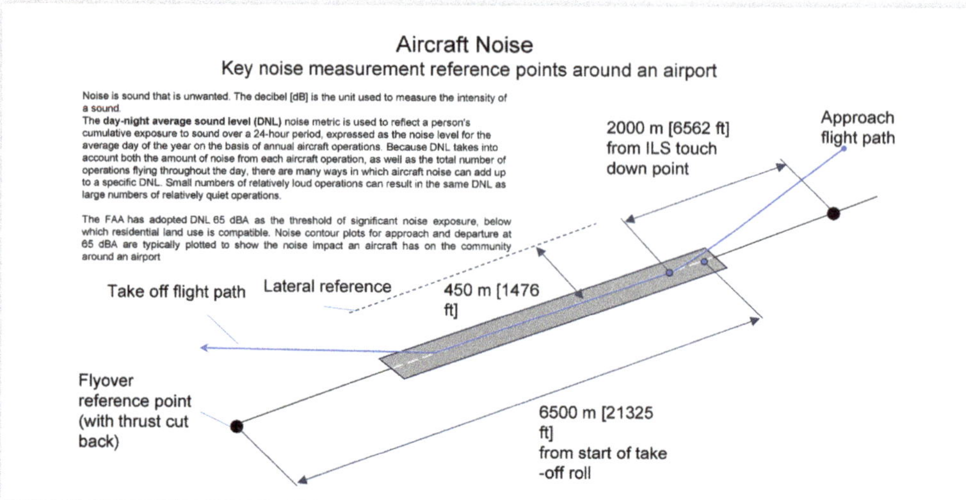

[1] https://www.faa.gov/documentLibrary/media/Advisory_Circular/AC_25_807-1.pdf.

During the time of this writing, the latest noise requirement by the ICAO, as outlined in Annex 16, Volume 1 Chapter 14, is the Chapter 14 noise requirement. In the US, the ICAO Chapter 14 ruling is defined by the FAA as Stage 5, and the rule has been effective since November 3, 2017.

The ruling is typically described in terms of the cumulative margin in effective perceived noise decibels (EPNdB) required relative to Chapter 3, which is the reference. Chapter 14 currently is defined as Chapter 3 + 17 EPNdB of cumulative margin. The figure shows how the cumulative level varies with maximum take-off weight.

FIGURE E.2 ICAO noise standard.[2]

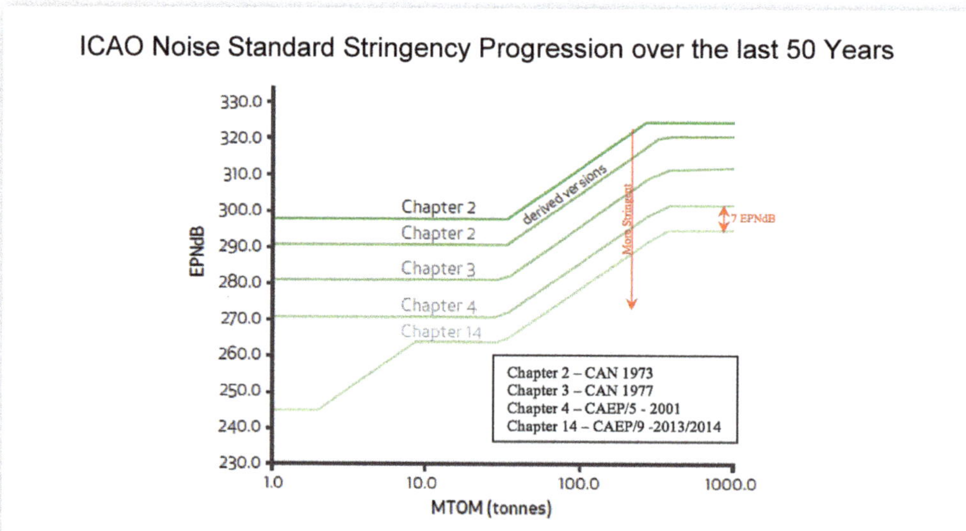

[2] https://www.icao.int/Meetings/EnvironmentalWorkshops/Documents/2015-Warsaw/3_2_Aircraft-Noise-Technology-and-International-Noise-Standards.pdf#search=ICAO%20Noise%20standard.

Quota Count (QC)—night-time aircraft movements at Heathrow, Gatwick, and Stansted airports are restricted by movement limits and noise quotas that are set by the Department for Transport (DfT). The noise quotas are designed to encourage the use of quieter aircraft. Movements at each airport count against the airport's noise quota according to their Quota Count (QC) classifications, which are based on ICAO-certificated noise levels.

The present Quota Count (QC) system, which was introduced in 1993, classifies aircraft into different categories depending on their ICAO noise certification data. Certificated noise levels are measured in effective perceived noise decibels (EPNdB), a specialized noise unit used for aircraft noise certification tests and are referred to as effective perceived noise levels (EPNLs).

Different types of aircraft (based on airframe, engine type, and maximum take-off or landing weight) are classified separately for landing and take-off on the basis of their certificated noise levels into one of the following QC categories (Table E.1).

TABLE E.1 Noise classification and the quota count.

Noise classification, EPNdB	Quota count
Below 81	0
81–83.9	0.125
84–86.9	0.25
87–89.9	0.5
90–92.9	1.0
93–95.9	2.0
96–98.9	4.0
99–101.9	8.0
Greater than 101.9	16.0

Aircraft noise is typically made up of two components, typically the airframe noise and jet engine noise. The airframe noise sources can be caused by deployed flaps as the air whistles through the various gaps in the high lift system, and the deployed landing gear. Jet engine noise is just the noise emanating from the turbo machinery at the nozzle and the inlet of the aircraft.

There is an interplay between the airframe and engine noise. During departure, the engine noise typically dominates the noise signature of the aircraft due to the high thrust required for take-off. In the take-off phase, high-lift devices are deployed for a high lift-to-drag ratio and are not at the maximum deflection angles. Low drag typically translates into low noise. Also, during take-off, the landing gear is retracted soon after liftoff, further reducing the exposure of the airframe as a dominant noise source. A combination of high lift-to-drag ratio wings and high-bypass ratio engines usually results in a lower noise signature. While a higher thrust is not favorable for obvious reasons, a higher thrust-to-weight ratio does allow the aircraft to climb at a higher rate, and this can position the aircraft at a higher altitude over the microphones, measuring noise at the departure end of the runway, and this can have an appreciable effect on the noise signature of the aircraft heard on the ground.

Like everything in airplane design, optimizing for noise is a balance between a high lift-to-drag ratio, engine bypass ratio, and engine thrust.

On approach to landing, the tables are turned typically. Due to the low-thrust setting, the noise signature of the aircraft is dominated by the airframe. This is primarily due to the high-lift devices being set at their maximum lift settings, which also translates into higher drag. The landing gear also plays a more prominent role in approach noise as it is deployed and hanging out in the wind. Open wheel wells (such as on the B737) can exacerbate the noise signature of the airframe. Here too, a combination of a high-bypass ratio engine and high lift-to-drag ratio wing help reduce the noise signature. A high lift-to-drag ratio translates into a lower thrust setting on approach. While airframe noise typically dominates, the noise from the engine cannot be ignored. Lower bypass ratio engines can, in some circumstances, also be a dominant source of noise. Specifics vary from aircraft to aircraft, depending on the engine airframe combination.

Wake Vortex

Wake vortex is defined as the turbulence that is generated by the passage of an aircraft in flight. It includes various components, the most important of which are wingtip vortices as they can persist for up to three minutes after an aircraft passes. Wingtip vortices occur when a wing is generating lift, and they can be hazardous, especially during the take-off and landing phases of flight.

The vortices from larger aircraft pose problems to encountering aircraft. For instance, the wake of these aircraft can impose rolling moments exceeding the roll-control authority of the encountering aircraft. This then leads to separation distances determined by certification authorities depending on the leading and trailing aircraft pair.

The strength of the vortex is directly proportional to the aircraft weight and inversely proportional to the air density, aircraft velocity, and the aircraft's aerodynamic wingspan. The vortex strength can be written in equation form as follows:

$$G = \left\{ W / \left[r \times U \times \left(0.77 \times b_{aero} \right) \right] \right\} \tag{E.1}$$

where

G is the vortex strength and is also called the circulation

W is the aircraft weight

ρ is the air density

U is the aircraft speed

0.77 factor reflects the span of the vortex tips based on a typical wing efficiency of 0.975 in the CLEAN configuration (flaps and slats retracted)

b_{aero} is the aerodynamic wingspan

Typically, the greatest vortex strength occurs when the aircraft generating the vortex is heavy, slow, and clean. Depending on the high-lift configuration, the flaps can aid in reducing vortex strength.

Just as important as the strength of the vortex is the descent speed of the vortex. The main interest in the study of wake turbulence is in keeping the trailing aircraft separated from the vortex-generating aircraft in the take-off and landing flight path. Typically, the stronger the vortex, the more quickly it descends to the ground and dissipates. Un-intuitively, a weaker vortex lingers longer in the same spot and can be equally hazardous to the trailing aircraft following it.

The vortex descent velocity can be written in equation form as follows:

$$V_{desc} = G / \left[2 \times P \times 0.77 \times b_{aero}\right] \tag{E.2}$$

where
V_{desc} is the vortex descent velocity
G is the vortex strength and is also called the circulation
0.77 factor reflects the span of the vortex tips based on a typical wing efficiency of 0.975 in the clean configuration (flaps and slats retracted)
b_{aero} is the aerodynamic wingspan

Hence, a heavy aircraft with a smaller wingspan will have a vortex that will descend faster and dissipate faster.

Separations and the ICAO Rules [85]

Almost everywhere in the world, the separations comply with the ICAO rules.

Classifications: Three categories of aircraft are defined according to the MTOW:

- Heavy (H): above 136 tons (299,880 lb)
- Medium (M): between 7 (15,435 lb) and 136 tons (299,880 lb)
- Light (L): below 7 tons (15,435 lb)

In addition, despite being classified as heavy, the A380 is known as super (S) and subjected to increased separations in approach, behind.

In cruise, the separations are identical for all aircraft types:

- Horizontally: 5 NMI
- Vertically: 1000 ft

On approach, the separations depend on the leader and the follower classification. Table E.2 gives the separations for the various pairs on the same runway. They apply also to operations on different parallel runways if they are separated laterally by less than 760 m (2493 ft).

Note: The A380 separations are not in the ICAO recommendations (PANS-ATM), but in a provisional State Letter published by ICAO in 2008.

TABLE E.2 ICAO separation distances on approach to landing in nautical miles.

	Aircraft following			
Aircraft leading	**Super heavy**	**Heavy**	**Medium**	**Light**
Super heavy	—	6 NMI	7 NMI	8 NMI
Heavy and B757	—	4 NMI	5 NMI	6 NMI
Medium	—	—	—	5 NMI
Light	—	—	—	—

Note: Boeing 757 wake characteristic is similar to that of a heavy transport.

Emissions

There are primarily two main areas of focus with regard to emissions as they relate to transport aircraft: carbon dioxide (CO_2) emissions and nitrogen oxide (NOx) emissions. CO_2 emissions corelate directly to the fuel burn of the aircraft, and international aviation has seen a decline in CO_2 emissions due to the steady introduction of new jet transports since the mid-2000s, such as the Boeing 777-300ER, 787, 747-8, and 737 MAX, and the Airbus A350, A320neo, and A330 neo family. Engines are the prime source of NOx emissions, which have also seen a decrease due to more efficient engine technologies that have accompanied the new jet transports.

Current industry understanding is that aviation emissions of CO_2 are less than 3% of the global greenhouse emissions. The amount of CO_2 emissions from aviation is expected to grow around 3–4% per year. Medium-term mitigation for CO_2 emissions from the aviation sector can and will potentially come from improved fuel efficiency. However, such improvements are expected to offset the growth of CO_2 aviation emissions only partially.

ICAO long-term aspirational goals for the international aviation sector are a 2% annual fuel efficiency improvement through 2050 and carbon neutral growth from 2020 onward. This is to be achieved by various means including aircraft technology improvements (airframe, engine), operational technology improvements (direct descent, direct routing, etc.), sustainable aviation fuels, and market-based measures such as the Carbon Offsetting and Reduction Scheme for International Aviation (CORSIA).

In 2017, the 36-State ICAO Council adopted a new aircraft CO_2 emissions standard, which aims to reduce the impact of aviation greenhouse gas emissions on the global climate. The aircraft CO_2 emissions measure represents the world's first global design certification standard governing CO_2 emissions for any industry sector.

The standard will apply to new aircraft type designs from 2020 and to aircraft type designs already in-production as of 2023. Those in-production aircraft which, by 2028, do not meet the standard will no longer be able to be produced, unless their designs are sufficiently modified.

The 2028 ICAO CO_2 Standard [77, 84]

To establish the fuel efficiency of the airplane, the CO_2 metric system uses multiple test points to represent the fuel burn performance of an airplane type during the cruise phase of flight. Specifically, there are three averaged (i.e., equally weighted) points representing the airplanes high, middle, and low gross masses, which are calculated as a function of maximum take-off mass (MTOM). Each of these represents an airplane cruise gross mass seen regularly in service.

The objective of using three gross mass cruise points is to make the evaluation of fuel burn performance more relevant to day-to-day airplane operations. The metric system is based on the inverse of specific air range (i.e., 1/SAR), where SAR represents the distance an airplane travels in the cruise flight phase per unit of fuel consumed. In some airplane designs, there are instances where changes in airplane size may not reflect changes in airplane weight, for example, when an airplane is a stretched version of an existing airplane design (e.g., the Boeing 787-9, which is stretched from the B787-8, or an Airbus A350-1000, which is stretched from the A350-900).

Note: To better account for such instances, not to mention the wide variety of airplane types and the technologies they employ, an adjustment factor was used to represent airplane size.

This is defined as the reference geometric factor (RGF), and it is a measure of airplane cabin size based on a two-dimensional projection of the cabin. This allows the CO_2 metric to better account for different airplane type designs. The overall capabilities of the airplane design is represented in the CO_2 metric system by the certified MTOM (or MTOW as used in the US). This accounts for many airplane design features, which allow it to meet market demand.

FIGURE E.3 The ICAO 2028 CO_2 standard.[3]

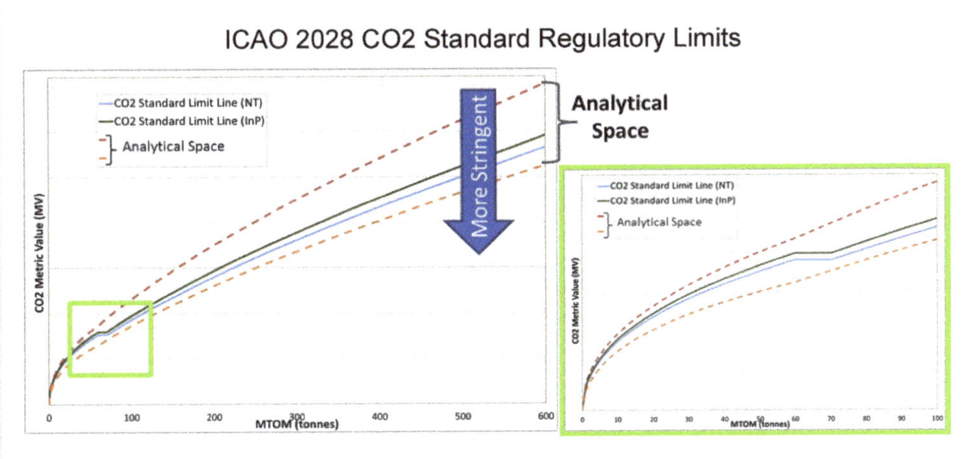

Courtesy of ICAO.

[3] https://www.icao.int/Meetings/EnvironmentalWorkshops/Documents/2015-Warsaw/3_2_Aircraft-Noise-Technology-and-International-Noise-Standards.pdf#search=ICAO%20Noise%20standard.

FIGURE E.4a The reference geometric factor.[4]

ICAO 2028 CO2 Standard Definitions

Airplane Maximum Take-Off Mass (MTOM) or MTOW

The highest possible take-off mass or weight for the type design configuration.
Three certification points are considered and equally weighted to derive a single metric system value.
The three certification test points evaluated are as follows:

- High gross weight = 0.92 x MTOW
- Low gross weight = [0.45 x MTOW] + [0.63 x MTOW $^{0.924}$]
- Mid gross weight = Average of high and low gross weight

Specific Air Range (1/SAR)

Specific Air Range is the distance an airplane travels in the cruise flight phase per unit of fuel consumed.

Reference Geometry Factor (RGF)

Reference Geometric Factor is an adjustment factor based on a measurement of aeroplane fuselage size derived from a two- dimensional projection of the fuselage.
The cabin area measured is at the *maximum* internal cabin width. This may or may not be the same width measured at floor level. On double deck aircraft such as the B747 or A380, the maximum cabin width is measured within each cabin (Main Deck and Upper Deck).

The cabin area includes aisles, assist spaces, passageways and stairwells (B747, A380). The cabin area excludes the rear pressure dome bulkhead which typically houses systems and lies outside of the utilized cabin space, crew rest or work areas which are not on the main deck, and permanent integrated fuel tanks within the cabin and above passenger floor

Courtesy of ICAO.

[4] https://www.icao.int/Meetings/EnvironmentalWorkshops/Documents/2015-Warsaw/3_2_Aircraft-Noise-Technology-and-International-Noise-Standards.pdf#search=ICAO%20Noise%20standard.

FIGURE E.4b

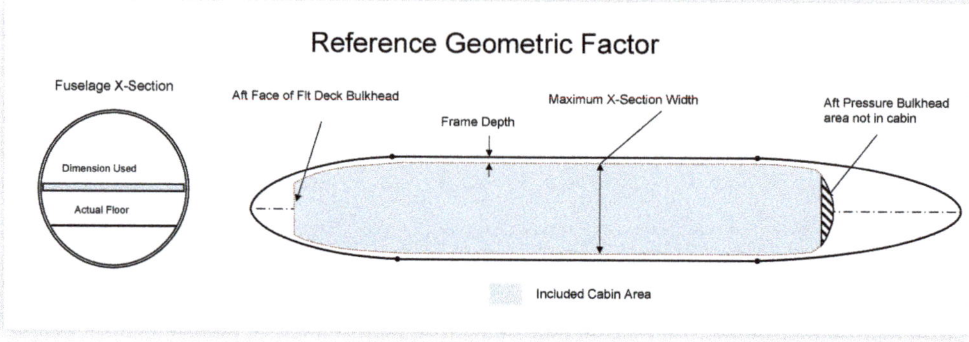
Reference Geometric Factor
Fuselage X-Section
Dimension Used
Actual Floor
Aft Face of Flt Deck Bulkhead
Frame Depth
Maximum X-Section Width
Aft Pressure Bulkhead area not in cabin
Included Cabin Area

Markus Mainka/Shutterst

Appendix F: Freighters

A critical part of the global trading system is the global value chain with production processes located across different countries. The global value chain enables companies to specialize in component production, research and development, or assembly. These specialized tasks originating in different countries are then combined through a complex network of trade links to produce finished goods like cars, cellular phones, and aircraft.

Air cargo is key to supporting the current global trading system. While air cargo transports less than 1% of goods by volume, an estimated 35% of the value of global trade is carried by air.

Freighter aircraft are a subcategory of jet transport that are dedicated to carrying cargo goods by air and they deserve special attention. Freighters have traditionally followed their passenger variants as a member of an aircraft family, but their success depends a lot on the foresight of the aircraft designer to plan for this specialized member of the aircraft family. Much like a passenger jet, the success of a freighter depends upon its versatility in both payload and range carrying capability.

Key Vocabulary for Freighter Aircraft

New Build Freighter

A factory built, or new build freighter is typically based on a passenger variant or otherwise. Examples include the Boeing 767-300F, 747-8F, 777F, and the Airbus A330-200F. Those in development include the Boeing 777-8F and Airbus A350F. These vehicles are designed, developed, built, and delivered from the factory by the aircraft manufacturer.

New build freighters are typically designed from the ground up to be more capable and have dedicated airframe architectures (floor, fuselage panels, wing center sections, wing high lift devices, etc.) optimized to carry heavy cargo loads. While more expensive, they are also preferred by operators for their payload-range capability and availability in revenue service. This is especially true of express carriers such as FedEx, DHL, and UPS.

So far, new build freighters have been limited to the wide body market, and while there has been a lot of interest in building new narrow body freighters, the life cycle cost of acquisition and operation have still to be favorable enough to warrant a product in this space.

Converted Freighter

A converted freighter is an aircraft that has spent most of its service life as a passenger variant and then converted to a freighter by the manufacturer or a maintenance conversion facility. Converted freighters are typically less expensive but can sometimes come at the expense of reliability because they are based on used passenger airframes that are 15 to 20 years old, if not more. Hence, building a fleet of aircraft from converted passenger variants can mean variability in commodities like engine types, and flight deck differences among other variables common in a passenger fleet spread across multiple carriers. The term feedstock is often used to describe appropriately priced passenger airplanes that have reached an age (typically 20 years) where they have depreciated in value enough to make a good business case for a converted freighter. The business case accounts for the acquisition cost and the conversion cost to modify the used passenger aircraft with the necessary changes and engine overhauls to give the converted freighter a useful payload and utilization life beyond the completed operations in passenger service.

There are both wide- and narrow-body freighters in the converted freighter category. Exclusively, all narrow body freighters are converted freighters today.

The effect of the recent pandemic, along with market dynamics, has accelerated the availability of feedstock for conversions, and during the time of writing of this book, several widebodies such as the A330-300, 777-300ER, and the A321ceo are expected to be available at appealing price points. An estimate of conversion costs (in millions of dollars) based on gross payload capability in metric tons is given by Equation (F.1):

$$\text{Conversion cost} = \{0.35 \times \text{Gross payload} - 4.0\} \times 1.011^{(\text{YOC}-2022)} \quad \text{(F.1)}$$

where
YOC is the year of conversion
Gross payload is the gross structural payload in tons (including tare weight)

$$\text{Total cost} = \text{Acquisition cost} + \text{Conversion cost} \quad \text{(F.2)}$$

In 2022, it was estimated that with fuel prices rising and the advent of more fuel-efficient aircraft such as the Boeing 787 and the A330neo, there would be significant feedstock of older A330-300 aircraft being phased out of passenger service and into the conversion market space at affordable prices.

Assuming the average price of a 15-year-old A330-300 to be \$15 million (in 2023) and a conversion cost for the 63-metric ton freighter to be \$18.25 million, then the total cost of acquiring a passenger A330-300 and converting it to a freighter in 2023 would be \$33.25 million.

Looking at published 2022 pricing for new build freighters from Airbus and Boeing, the list price or acquisition cost of a new build freighter can be estimated as follows:

$$\text{List price}(\$\text{M}) = \{6.51 \times 10^{-7} \times \text{Productivity}\} \times 1.011^{[\text{EIS}-2022]} \quad \text{(F.3)}$$

where

$$\text{Productivity} = \left[\text{Revenue payload} \times \text{Range}^{0.85}\right] / \left[\text{Mission fuel / Revenue payload}\right]^{0.85} \quad \text{(F.4)}$$

Revenue payload = Gross payload – Tare weight
Mission fuel is the total fuel including reserve fuel for the design mission in pounds
Range is the design range in nautical miles
EIS is the entry into service date

Note: As stated in Chapter 2, typical prices can be at deep discounts in the neighborhood of 50% or more off the list price.

Revenue Payload

Revenue payload is the revenue generating cargo payload and is usually expressed in pounds or kilograms. The Boeing Company typically quotes revenue payload in their marketing material. In Boeing's case, the tare weight is included in the operating empty weight of the aircraft.

Gross Structural Payload

Gross payload is the sum of the revenue payload and tare weight. Airbus typically quotes gross structural payload in their marketing material. In Airbus aircraft, the tare weight is not included in the operating empty weight of the aircraft.

Cargo Volume

Cargo volume is the containerized and palletized volume on the aircraft. Typically, bulk cargo area volume in a wide body transport is not included in the numbers quoted by manufacturers, but this can vary.

Commercial freighters usually have their total volume distributed so that approximately 80% of the volume is on the main deck, and approximately 20% is on the lower decks (forward and aft). An inspection of existing commercial freighters utilizing **pallets** on the main deck and lower deck helps us estimate the main deck volume as follows:

$$\text{Volume}_{\text{main deck (cu-ft)}} = 37.2 \times \left[L_f \times Df_{eff}^{\ 0.45}\right] - 11{,}690 \quad \text{(F.5)}$$

where
L_f is the fuselage length in feet
Df_{eff} is the fuselage effective diameter = [Fuselage height × Fuselage width]$^{-0.5}$

$$\text{Volume}_{\text{total pallet (cu-ft)}} = 1.25 \times \text{Volume}_{\text{main deck (cu-ft)}} \quad \text{(F.6)}$$

Utilizing containers instead of pallets adds about 4.5% to the total palletized volume.

$$\text{Volume}_{\text{total containerized (cu-ft)}} =\sim 1.045 \times \text{Volume}_{\text{total pallet (cu-ft)}} \quad \text{(F.7)}$$

***Note*:** Freighter aircraft like the 747 where the flight deck is above the main deck can have a small increase in main deck volume. However, due to the presence of body gear, which takes up lower deck cargo volume, the total volume is reduced by approximately the same amount.

Cargo Density

Cargo density is defined as the revenue payload divided by the available cargo volume and is usually measured in pounds per cubic foot or kilograms per cubic meter.

Volumetric Payload

Revenue payload typically evaluated at a density of 7 lb/ft^3. The purpose of using this number is to evaluate the package freight carrying capability of the aircraft. It can highlight the efficiency of a freighter at average payloads typically seen in day-to-day operations.

Unit Load Device or ULD

This is the physical pallet or container type of some dimension that is used to carry the payload.

Pallet

Typically, a metal plate of some dimension that the payload is set upon and secured via a cargo net or tie down apparatus.

Container

A closed device of some dimension that carries the payload.

Tare Weight

The weight of the empty ULD. The term usually refers to the total weight of the mix of devices used to carry the payload. The term "tare" is often used interchangeably with "tare weight." Typically, containers weigh more than pallets. Tare weight can vary significantly depending on the type of ULD device and the mix of pallets and containers.

There is no fixed standard when one is talking about tare weight. However, when considering the design of a freighter transport, it is important to reasonably estimate a tare weight in the early stages of a design. An estimate of tare weight, **in pounds**, for pallets and containers is given as follows:

$$\text{Total tare weight}\left(\text{all pallets}\right) = 0.494 \times \text{Volume}_{\text{total pallet}\,(\text{cu-ft})} \tag{F.8}$$

$$\text{Total tare weight}\left(\text{all containers}\right) = 0.945 \times \text{Volume}_{\text{total containerized}\,(\text{cu-ft})} \tag{F.9}$$

***Note*:** Volumes quoted in the aforementioned equations (in ft^3) do not include bulk cargo volume

Cargo Handling System or CHS

The system of rollers, locks, power drive units (in the case of a powered cargo handling system), and guide rails make up the floor of the cargo deck. A cargo handling system can be a manual system or powered system with a series of powered drive units which are controlled by a cargo control unit. Manual systems are lighter than powered systems and are typically found on "medium freighters" that carry less than 80 tons of revenue payload. Manual systems also typically require the aircraft to have a relatively flat attitude on the ground. Ideally less than degree, to allow easy loading and offloading of cargo by ground crew.

General Market Freighter

A freighter aircraft capable of carrying heavy and outsized cargo with average cargo densities of 9.0 lb/ft^3 or higher. The 777F and 747-8F are a good example of a capable general market freighter, although the 767-300F is also used successfully to carry general market freight densities. Atlas Air, Qatar Airways, and Cargolux are a good example of general market freight operators.

Package Freighter

A freighter aircraft capable of carrying "lighter freight" with an average cargo density of 7 lb/ft^3. The 767-300F and A300-600F are a good example of capable package freighters, although both the 777F and 747-8F are also used to carry package freight very successfully. DHL, FedEx, and UPS are good examples of package freight operators.

When payload and range numbers for a freighter aircraft are quoted, they are typically done so at the corner point which depicts the intersection of the maximum zero fuel weight (MZFW) and the maximum take-off weight (MTOW) on a payload-range diagram. Capable intercontinental freighters are sized such that the maximum revenue payload can be carried about 4450 NMI, which is representative of a Hong Kong to Anchorage (Alaska) mission. Anchorage is a major freighter hub and is used as a technical stop that connects Asia to the North American market.

Freight Aircraft Considerations

Typically, aircraft manufacturers do not design an aircraft family specifically around the requirements of the freighter family member. The reason being that a freighter is usually the last of the family of aircraft being designed and the designer can utilize available margin in the airframe and engine to maximize the payload and range capability of the design.

Having said that, the most important consideration for a freighter, even in the early design phase of the passenger variant, is the fuselage cross-sectional design. Every effort should be made to see if existing pallets and containers can be accommodated within the new cross section. While a penalty in aircraft weight and drag for a passenger variant is not desired, a small increase in fuselage cross section to enable a robust freighter further down the road usually pays dividends in the long run.

A freighter aircraft family member should be considered early in the design phase of a passenger transport because these aircraft can drive several design decisions as follows:

- Fuselage cross-sectional height and width.
  - To dimensionally accommodate unit load devices available in the market.
  - To dimensionally accommodate a large main deck door and door mechanism.
  - If possible, accommodate a reasonable walkway on either side of the main deck, which is a preference on a large freighter.

- Wing structural design.
- High lift system.
  - Higher zero fuel weights and maximum landing weights than the passenger variant.
  - Higher take-off weight potential than the passenger variant.
- Air conditioning system that accommodates temperature-sensitive cargo.
  - Pharmaceuticals and live animal carriage.
- More capable brakes to accommodate daily operations.
  - Higher maximum landing weight.
  - Higher approach speeds.
- Improved load ability to accommodate large variations in cargo weight.
- Main deck cargo door location to provide flexibility for ground loading operations.
- A reasonable wheelbase (distance between nose wheel and main wheels).
  - Accommodate operations in remote airfields with narrow and less load capable taxiways and runways.
  - Improved turning radius may be required on narrow pavement.
- Fire suppression on the main deck.
  - Typically done today via forced decompression to starve the fire of oxygen.
  - The advent of lithium-ion batteries and their widespread proliferation into electronic devices requires the freight industry to put more emphasis on different fire suppression methods to combat the extreme intensity of lithium ion battery fires.

Quick Rules of Thumb

Once a passenger aircraft operating empty weight is known, a quick estimate can be made to estimate a freighter OEW.

$$OEW_{\text{Freighter with tare weight included}} = OEW_{\text{passenger}} \times \sim 0.92 \tag{F.10}$$

Here, the 0.92 factor is used to estimate the following:

- Removal of passenger accommodations.
- Removal all lavatories except one for use by the crew.
- Removal and replacement of passenger floor with a more capable freighter floor.

- Reduction of the potable water and waste tank capacity for crew and supernumerary.
- Removal of all passenger doors except Door 1 (left and right) for crew access.
- Addition of a basic supernumerary area for cargo handlers, live animal care takers, and pilot rest areas.
- Addition of a large main deck cargo door (MDCD) for freighter operations.
- Removal of the passenger windows and re-paneling of the fuselage skins for the higher maximum zero fuel weights.
- Strengthening of the wing center section to carry the heavier floor loads over the wing.
- Additional airframe strengthening due to higher design weights and approach speeds (sizes flap support architecture).

Note: For airplanes like the 747, when a new-build factory freighter is designed, the upper deck is shortened to reduce unnecessary airframe structural weight. In this case, a factor may be more along the lines of 0.875 due to the removal of the large part of the upper deck.

FIGURE F.1a B747-8 passenger with extended upper deck.

FIGURE F.1b B747-8F with shortened upper deck.

Ryan Fletcher/Shutterstock.com.

The operating empty weight quoted earlier includes the tare weight and is the Boeing standard for quoting freighter operating empty weights. The Airbus standard, on the other hand, includes the tare weight as part of the payload (or rather gross payload), and Airbus OEWs typically may appear lighter due to the missing tare weight.

$$\text{OEW}_{\text{Freighter without tare}} = \text{OEW}_{\text{Freighter with tare weight included}} - \text{Tare weight} \tag{F.11}$$

The manufacturer's empty weight (MEW) can be estimated as follows:

$$\text{MEW}_{\text{Freighter}} = \sim 0.975 \times \text{OEW}_{\text{Freighter without tare}} \tag{F.12}$$

A Promising Future for Air Cargo and Freighter Transports

Pre-pandemic, roughly 50% of air cargo was carried in the belly of passenger jets and 50% by dedicated freighter aircraft. During the peak of the pandemic in 2020, with the suspension of passenger flights, almost all cargo transported by air was done so on dedicated freighters. As of April of 2023, dedicated freighters were still carrying over 65% of the world's air freight. This number is expected to decline, once passenger air service (and hence belly cargo on these flights) fully recovers.

However, it is forecasted that cargo yields will likely stabilize at a higher $2.0 per kilogram post-pandemic versus the $0.8 per kilogram pre-pandemic level.

The medium- to long-term market outlook for freighter aircraft remains strong. One can say with a fair bit of certainty that the air cargo market will see sustained growth, which is currently forecasted to be on the order of 4% (between 2021 and 2041). This growth is supported by several factors, such as GDP growth, rising container shipping prices, ongoing global supply chain constraints, and strong growth in e-commerce globally.

Boeing predicts the global fleet of freighters will grow, with 940 production freighters delivered by 2041 and over 1850 freighter conversion deliveries by the same period. Recently launched products like the Boeing 777-8F and the Airbus A350F are the next generation of large cargo aircraft on the drawing board to meet the growing demand for air freight.

Both of these products also promise to be more sustainable, with significantly lower carbon emissions that meet the newly introduced ICAO CO_2 standards. Lower carbon emissions are a direct result of improved fuel efficiency, which, in turn, means a lower cost of doing business and more affordable goods for everyone.

Air freight is a key contributor to the part of global commerce that requires high value, just in time and perishable goods that people and industry depend upon. As technologies continue to improve aircraft capability, it is likely this part of the aviation industry will continue to see growth.

Index

About the Author

© Zarir D. Pastakia.

Zarir D. Pastakia is an Indian American with over 23 years of industry experience in commercial aircraft design, competitive analysis, and business strategy.

Born and brought up in an airline family, his love for airplanes started at a very young age, and learning about aircraft has been and continues to be a lifelong passion. One can never learn enough in this dynamic and forever evolving industry. He has been fortunate enough to work with talented aircraft designers from Boeing, Airbus, and Fairchild Dornier.

Zarir believes in giving back to the aviation community by helping to mentor and develop the next generation of aircraft designers.

Most beautiful aircraft of all time: Supermarine Spitfire

Favorite aircraft of all time: Boeing 777

www.ingramcontent.com/pod-product-compliance
Lightning Source LLC
LaVergne TN
LVHW071342210926
848486LV00003B/207